Renewable Energy

Renewable Energy

Michael Silva

SYRAWOOD
PUBLISHING HOUSE

New York

Published by Syrawood Publishing House,
750 Third Avenue, 9th Floor,
New York, NY 10017, USA
www.syrawoodpublishinghouse.com

Renewable Energy
Michael Silva

International Standard Book Number: 978-1-64740-043-9 (Hardback)

Cataloging-in-Publication Data

Renewable energy / Michael Silva.
    p. cm.
Includes bibliographical references and index.
ISBN 978-1-64740-043-9
1. Renewable energy sources. 2. Power resources.
3. Renewable natural resources. I. Silva, Michael.
TJ808 .R46 2020
333.794--dc23

Table of Contents

Preface

This book has been written, keeping in view that students want more practical information. Thus, my aim has been to make it as comprehensive as possible for the readers. I would like to extend my thanks to my family and co-workers for their knowledge, support and encouragement all along.

Renewable energy, also known as green energy, is obtained from renewable resources. Such natural resources include sunlight, rain, waves, geothermal heat and wind. It plays an important role in electricity generation, transportation, air and water heating or cooling, and off-grid energy services. Some of the different technologies which are used to produce renewable energy are wind turbines for converting wind power into electricity, water turbines for producing hydroelectricity and photovoltaic cells for harnessing solar energy. Various other renewable technologies, such as cellulosic ethanol, marine energy and hot-dry-rock geothermal power, are also being developed. There are numerous benefits of using renewable energy over fossil fuels such as reducing air pollution and improving public health. Different approaches, evaluations and methodologies and advanced studies on renewable energy have been included in this book. Such selected concepts that redefine this area have been presented herein. Those in search of information to further their knowledge will be greatly assisted by this book.

A brief description of the chapters is provided below for further understanding:

Chapter – What is Renewable Energy?

The energy which is derived from renewable resources, such as wind, rain, sunlight, tides, waves and geothermal heat, is referred to as renewable energy. This is an introductory chapter which will introduce briefly all the significant aspects of renewable energy such as its advantages, disadvantages and economics.

Chapter – Solar Energy and Photovoltaic Systems

The radiation from the sun that is capable of producing heat by causing chemical reactions is referred to as solar energy. Some of the important aspects of solar energy are solar power and solar thermal energy. This chapter discusses in detail these aspects related to solar energy as well as solar panels and photovoltaic systems.

Chapter – Understanding Wind Energy

The utilization of air flow with the help of wind turbines to provide mechanical power is referred to as wind energy. The production of power and electricity through wind energy takes place at wind farms. This chapter has been carefully written to provide an easy understanding of the varied facets of wind energy as well as its advantages and disadvantages.

Chapter – Geothermal Energy and Technologies

The heat which comes from the sub-surface of the Earth is known as geothermal energy. Various geothermal technologies are used to harness the Earth's heat. The main types of such technologies are ground source heat pumps, direct use geothermal and deep and enhanced geothermal systems. These applications of geothermal energy have been thoroughly discussed in this chapter.

Chapter – Biomass Energy

The renewable energy which is derived from the organic material that is obtained from plants and animals is known as biomass energy. Biofuel is the most common fuel that is derived from biomass. This chapter closely examines the varied uses of biomass energy to provide an extensive understanding of the subject.

Chapter – Environmenmental Impacts of Renewable Energy

The impact of renewable energy upon the environment depends on numerous factors such as the technologies used and geographic locations. The environmental impacts of various renewable sources of energy such as solar energy, wind power, hydroelectric power and geothermal power have been thoroughly discussed in this chapter.

Michael Silva

1

What is Renewable Energy?

The energy which is derived from renewable resources, such as wind, rain, sunlight, tides, waves and geothermal heat, is referred to as renewable energy. This is an introductory chapter which will introduce briefly all the significant aspects of renewable energy such as its advantages, disadvantages and economics.

Renewable energy, often referred to as clean energy, comes from natural sources or processes that are constantly replenished. For example, sunlight or wind keep shining and blowing, even if their availability depends on time and weather.

While renewable energy is often thought of as a new technology, harnessing nature's power has long been used for heating, transportation, lighting, and more. Wind has powered boats to sail the seas and windmills to grind grain. The sun has provided warmth during the day and helped kindle fires to last into the evening. But over the past 500 years or so, humans increasingly turned to cheaper, dirtier energy sources such as coal and fracked gas.

Now that we have increasingly innovative and less-expensive ways to capture and retain wind and solar energy, renewables are becoming a more important power source, accounting for more than one-eighth of U.S. generation. The expansion in renewables is also happening at scales large and small, from rooftop solar panels on homes that can sell power back to the grid to giant offshore wind farms. Even some entire rural communities rely on renewable energy for heating and lighting.

Dirty Energy

Nonrenewable, or "dirty," energy includes fossil fuels such as oil, gas, and coal. Nonrenewable sources of energy are only available in limited amounts and take a long time to replenish. When we pump gas at the station, we're using a finite resource refined from crude oil that's been around since prehistoric times.

Nonrenewable energy sources are also typically found in specific parts of the world, making them more plentiful in some nations than others. By contrast, every country has access to sunshine and wind. Prioritizing nonrenewable energy can also improve national security by reducing a country's reliance on exports from fossil fuel–rich nations.

Many nonrenewable energy sources can endanger the environment or human health. For example, oil drilling might require strip-mining Canada's boreal forest, the technology associated with fracking can cause earthquakes and water pollution, and coal power plants foul the air. To top it off, all these activities contribute to global warming.

Types of Renewable Energy Sources

Solar Energy

Humans have been harnessing solar energy for thousands of years—to grow crops, stay warm, and dry foods, "more energy from the sun falls on the earth in one hour than is used by everyone in the world in one year." Today, we use the sun's rays in many ways—to heat homes and businesses, to warm water, or power devices.

Solar panels on the rooftops of East Austin, Texas.

Solar, or photovoltaic (PV), cells are made from silicon or other materials that transform sunlight directly into electricity. Distributed solar systems generate electricity locally for homes and businesses, either through rooftop panels or community projects that power entire neighborhoods. Solar farms can generate power for thousands of homes, using mirrors to concentrate sunlight across acres of solar cells. Floating solar farms—or "floatovoltaics"—can be an effective use of wastewater facilities and bodies of water that aren't ecologically sensitive.

Solar supplies a little more than 1 percent of U.S. electricity generation. But nearly a third of all new generating capacity came from solar in 2017, second only to natural gas.

Solar energy systems don't produce air pollutants or greenhouse gases, and as long as they are responsibly sited, most solar panels have few environmental impacts beyond the manufacturing process.

Wind Energy

Today, turbines as tall as skyscrapers—with turbines nearly as wide in diameter—stand at attention around the world. Wind energy turns a turbine's blades, which feeds an electric generator and produces electricity.

Wind, which accounts for a little more than 6 percent of U.S. generation, has become the cheapest energy source in many parts of the country. Top wind power states include California, Texas, Oklahoma, Kansas, and Iowa, though turbines can be placed anywhere with high wind speeds—such as hilltops and open plains—or even offshore in open water.

Other Alternative Energy Sources

Hydroelectric Power

Hydropower is the largest renewable energy source for electricity in the United States, though wind energy is soon expected to take over the lead. Hydropower relies on water—typically fast-moving water in a large river or rapidly descending water from a high point—and converts the force of that water into electricity by spinning a generator's turbine blades.

Nationally and internationally, large hydroelectric plants—or mega-dams—are often considered to be nonrenewable energy. Mega-dams divert and reduce natural flows, restricting access for animal and human populations that rely on rivers. Small hydroelectric plants (an installed capacity below about 40 megawatts), carefully managed, do not tend to cause as much environmental damage, as they divert only a fraction of flow.

Biomass Energy

Biomass is organic material that comes from plants and animals, and includes crops, waste wood, and trees. When biomass is burned, the chemical energy is released as heat and can generate electricity with a steam turbine.

Biomass is often mistakenly described as a clean, renewable fuel and a greener alternative to coal and other fossil fuels for producing electricity. However, recent science shows that many forms of biomass—especially from forests—produce higher carbon emissions than fossil fuels. There are also negative consequences for biodiversity. Still, some forms of biomass energy could serve as a low-carbon option under the right circumstances. For example, sawdust and chips from sawmills that would otherwise quickly decompose and release carbon can be a low-carbon energy source.

Geothermal Energy

The Svartsengi geothermal power plant near Grindavík, Iceland.

If you've ever relaxed in a hot spring, you've used geothermal energy. The earth's core is about as hot as the sun's surface, due to the slow decay of radioactive particles in rocks at the center of the planet. Drilling deep wells brings very hot underground water to the surface as a hydrothermal resource, which is then pumped through a turbine to create electricity. Geothermal plants typically have low emissions if they pump the steam and water they use back into the reservoir. There are ways to create geothermal plants where there are not underground reservoirs, but there are concerns that they may increase the risk of an earthquake in areas already considered geological hot spots.

Ocean

Tidal and wave energy is still in a developmental phase, but the ocean will always be ruled by the moon's gravity, which makes harnessing its power an attractive option. Some tidal energy approaches may harm wildlife, such as tidal barrages, which work much like dams and are located in an ocean bay or lagoon. Like tidal power, wave power relies on dam-like structures or ocean floor–anchored devices on or just below the water's surface.

Renewable Energy in the Home

Solar Power

At a smaller scale, we can harness the sun's rays to power the whole house—whether through PV cell panels or passive solar home design. Passive solar homes are designed to welcome in the sun through south-facing windows and then retain the warmth through concrete, bricks, tiles, and other materials that store heat.

Some solar-powered homes generate more than enough electricity, allowing the homeowner to sell excess power back to the grid. Batteries are also an economically attractive way to store excess solar energy so that it can be used at night. Scientists are hard at work on new advances that blend form and function, such as solar skylights and roof shingles.

Geothermal Heat Pumps

Geothermal technology is a new take on a recognizable process—the coils at the back of your fridge are a mini heat pump, removing heat from the interior to keep foods fresh and cool. In a home, geothermal or geoexchange pumps use the constant temperature of the earth (a few feet below the surface) to cool homes in summer and warm houses in winter—and even to heat water.

Geothermal systems can be initially expensive to install but typically pay off within 10 years. They are also quieter, have fewer maintenance issues, and last longer than traditional air conditioners.

Advantages of Renewable Energy

Less Global Warming

Human activity is overloading our atmosphere with carbon dioxide and other global warming emissions. These gases act like a blanket, trapping heat. The result is a web of significant and harmful impacts, from stronger, more frequent storms, to drought, sea level rise, and extinction.

In the United States, about 29 percent of global warming emissions come from our electricity sector. Most of those emissions come from fossil fuels like coal and natural gas.

In contrast, most renewable energy sources produce little to no global warming emissions. Even when including "life cycle" emissions of clean energy (ie, the emissions from each stage of a technology's life—manufacturing, installation, operation, decommissioning), the global warming emissions associated with renewable energy are minimal.

The comparison becomes clear when you look at the numbers. Burning natural gas for electricity releases between 0.6 and 2 pounds of carbon dioxide equivalent per kilowatt-hour (CO_2E/kWh); coal emits between 1.4 and 3.6 pounds of CO_2E/kWh. Wind, on the other hand, is responsible for only 0.02 to 0.04 pounds of CO_2E/kWh on a life-cycle basis; solar 0.07 to 0.2; geothermal 0.1 to 0.2; and hydroelectric between 0.1 and 0.5.

Renewable electricity generation from biomass can have a wide range of global warming emissions depending on the resource and whether or not it is sustainably sourced and harvested.

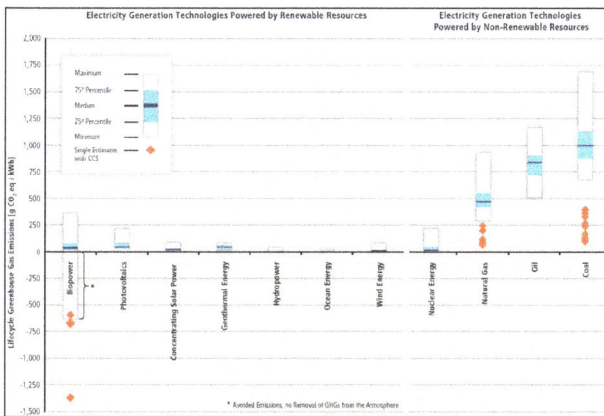

Different sources of energy produce different amounts of heat-trapping gases.
As shown in this chart, renewable energies tend to have much lower
emissions than other sources, such as natural gas or coal.

Increasing the supply of renewable energy would allow us to replace carbon-intensive energy sources and significantly reduce US global warming emissions. For example, a 2009 UCS analysis found that a 25 percent by 2025 national renewable electricity standard would lower power plant CO_2 emissions 277 million metric tons annually by 2025—the equivalent of the annual output from 70 typical (600 MW) new coal plants.

In addition, a ground-breaking study by the US Department of Energy's National Renewable Energy Laboratory (NREL) explored the feasibility of generating 80 percent of the country's electricity from renewable sources by 2050. They found that renewable energy could help reduce the electricity sector's emissions by approximately 81 percent.

Improved Public Health

The air and water pollution emitted by coal and natural gas plants is linked with breathing problems, neurological damage, heart attacks, cancer, premature death, and a host of other serious problems. The pollution affects everyone: one Harvard University study estimated the life cycle costs and public health effects of coal to be an estimated $74.6 billion every year. That's equivalent to 4.36 cents per kilowatt-hour of electricity produced—about one-third of the average electricity rate for a typical US home.

Most of these negative health impacts come from air and water pollution that clean energy technologies simply don't produce. Wind, solar, and hydroelectric systems generate electricity with no associated air pollution emissions. Geothermal and biomass systems emit some air pollutants, though total air emissions are generally much lower than those of coal- and natural gas-fired power plants.

In addition, wind and solar energy require essentially no water to operate and thus do not pollute water resources or strain supplies by competing with agriculture, drinking water, or other important water needs. In contrast, fossil fuels can have a significant impact on water resources: both coal mining and natural gas drilling can pollute sources of drinking water, and all thermal power plants, including those powered by coal, gas, and oil, withdraw and consume water for cooling.

Biomass and geothermal power plants, like coal- and natural gas-fired power plants, may require water for cooling. Hydroelectric power plants can disrupt river ecosystems both upstream and downstream from the dam. However, NREL's 80-percent-by-2050 renewable energy study, which included biomass and geothermal, found that total water consumption and withdrawal would decrease significantly in a future with high renewables.

Inexhaustible Energy

Strong winds, sunny skies, abundant plant matter, heat from the earth, and fast-moving water can each provide a vast and constantly replenished supply of energy. A relatively small fraction of US electricity currently comes from these sources, but that could change: studies have repeatedly shown that renewable energy can provide a significant share of future electricity needs, even after accounting for potential constraints.

In fact, a major government-sponsored study found that clean energy could contribute somewhere between three and 80 times its 2013 levels, depending on assumptions . And the previously mentioned NREL study found that renewable energy could comfortably provide up to 80 percent of US electricity by 2050.

Stable Energy Prices

Renewable energy is providing affordable electricity across the country right now, and can help stabilize energy prices in the future. Although renewable facilities require up-front investments to build, they can then operate at very low cost (for most clean energy technologies, the "fuel" is free). As a result, renewable energy prices can be very stable over time.

Moreover, the costs of renewable energy technologies have declined steadily, and are projected to drop even more. For example, the average price to install solar dropped more than 70 percent between 2010 and 2017 . The cost of generating electricity from wind dropped 66 percent between 2009 and 2016 . Costs will likely decline even further as markets mature and companies increasingly take advantage of economies of scale.

In contrast, fossil fuel prices can vary dramatically and are prone to substantial price swings. For example, there was a rapid increase in US coal prices due to rising global demand before 2008, then a rapid fall after 2008 when global demands declined. Likewise, natural gas prices have fluctuated greatly since 2000.

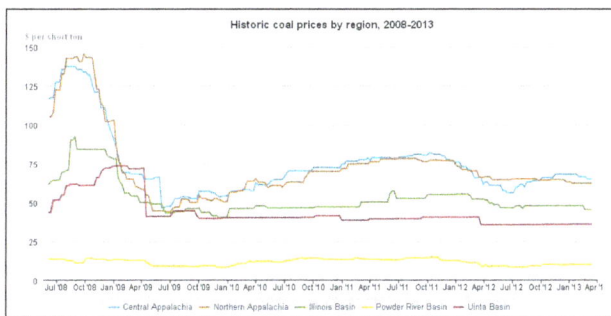

Historic coal prices by region, 2008-2013

Using more renewable energy can lower the prices of and demand for natural gas and coal by increasing competition and diversifying our energy supplies. And an increased reliance on renewable energy can help protect consumers when fossil fuel prices spike.

Reliability and Resilience

Wind and solar are less prone to large-scale failure because they are distributed and modular. Distributed systems are spread out over a large geographical area, so a severe weather event in one location will not cut off power to an entire region. Modular systems are composed of numerous individual wind turbines or solar arrays. Even if some of the equipment in the system is damaged, the rest can typically continue to operate.

For example, Hurricane Sandy damaged fossil fuel-dominated electric generation and distribution systems in New York and New Jersey and left millions of people without power. In contrast, renewable energy projects in the Northeast weathered Hurricane Sandy with minimal damage or disruption.

Water scarcity is another risk for non-renewable power plants. Coal, nuclear, and many natural gas plants depend on having sufficient water for cooling, which means that severe droughts and heat waves can put electricity generation at risk. Wind and solar photovoltaic systems do not require water to generate electricity and can operate reliably in conditions that may otherwise require closing a fossil fuel-powered plant.

The risk of disruptive events will also increase in the future as droughts, heat waves, more intense storms, and increasingly severe wildfires become more frequent due to global warming—increasing the need for resilient, clean technologies.

Disadvantages of Renewable Energy

1. It is not as Cost-Effective as other Energy Options

The costs of renewable energy have been going down. Some options are close to being competitive with non-renewable energy resources or have been subsidized to be competitive. From an industry standpoint, however, the lifetime costs of renewable energy are still nearly double that of what has been called conventional energy. The lifetime cost of coal, for example, is just 9.5 cents per kilowatt hour. The cost of off-shore wind, one of the cheapest renewable energy resources, is 15.8 cents per kilowatt hour.

2. It isn't Always a Commercially-Viable Option

Most renewable energy options must be collected at a specific geographical source. Imagine trying to collect high levels of solar energy while living in Seattle or using tidal energy while living in Nebraska. For renewable energy to be effective, it must have a distribution network created to transfer the energy to where it is needed. Those networks require non-renewable energies to be created, which offsets the benefits that the renewable energy generates for years, if not decades, after its installation.

3. It Still Generates Pollution

Renewable energy may be a better option for emission creation than fossil fuels, but that doesn't mean they are free from pollution. Many forms of renewable energy release particulates into the air. They may release carbon dioxide, or worse – methane. Part of this is due to the fact that the resources needed for renewables are built using fossil fuels, but not every renewable resource is clean. Biomass is a renewable energy and it burns organic matter directly into the atmosphere.

4. It may not be a Permanent Energy Resource

The environment evolves over time, shifting where renewable resources become available. Some locations, such as offshore wind or tidal energy, are generally quite reliable. Solar energy, however, can be difficult to predict. Geothermal energy may change over time. Billions of dollars can be spent to develop renewable energy resources only to have that money go to waste if that resource stops producing as expected.

5. It can often be Manipulated by Politics

Although renewable energy is generally accepted as the future of energy production globally, politics can be a negative influence on its development. If renewables are not given a political priority, then the industry tends to falter, and innovation is reduced in favor of non-renewable options. Because politics tends to go in cycles, renewables tend to see 4-8 years of growth, then 4-8 years of stagnation. The stops and starts make it difficult to create a thriving industry.

6. It is an Energy Resource that is Difficult to Access for many People

In the United States over a 30-year period, the government funded more than $100 billion in energy subsidies. Over half of those subsidies were directed toward nuclear power. Just 26% of those subsidies where funneled toward renewable energy technologies and energy efficiency priorities. Even with solar subsidies of up to 88 cents per kilowatt hour being provided, accessing that energy is not cost-effective for many low-income families. That means the priority stays on non-renewables, which further prevents innovation within this sector.

7. It can take a lot of Space to Install

Using current solar energy generation technologies, it takes over 40 hectares of panels to generate about 20 megawatts of energy. In comparison, a nuclear power plant of average size generates about 1,000 megawatts of energy on 259 hectares. If given the same amount of space, a solar energy facility would produce less than 200 megawatts. For land-based wind energy, a 2-megawatt turbine requires 1.5 acres of space. If given the same amount of space as a nuclear facility, even it would only generate a maximum of about 850 megawatts.

8. It isn't a Constant Energy Source

Then there is the issue with consistency. Nuclear and coal energy can be accessed at any time. Many renewables can only be collected during specific periods of the day. Solar energy, for example, isn't collected at night. Wind energy cannot be collected unless the wind is blowing. Even then, wind energy is often inefficient as the turbines have a minimum and maximum wind speed during which they operate.

9. It has Expensive Storage Costs

One of the pricing factors that is often excluded from the conversation on renewable energy is the storage cost. You must store the energy collected or you will lose it, which means having a battery installed. The overall storage cost for the energy is about 9 cents per kilowatt hour, but the cost of the battery is upfront. That means $10,000 to $25,000 upon installation just for the battery. Some types of batteries can wear out very quickly, especially if their full capacity is being used on a regular basis.

10. It has Large Capital Costs

If a new energy resource needs to be developed for a community right now, the capital cost of the resource will be a primary consideration. At this moment, the capital costs of non-renewable energy resources are lower than the capital costs of renewable energy. This may change in time, but is a primary reason why renewables are not favored in some parts of the world today.

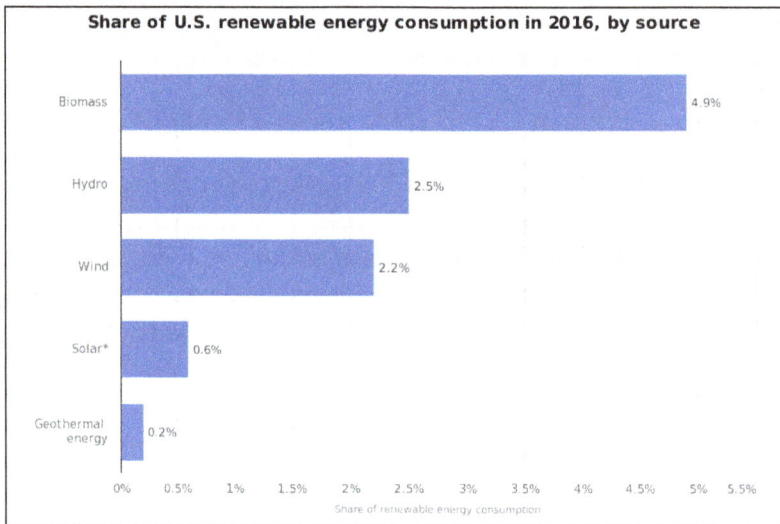

Share of U.S. renewable energy consumption in 2016, by source

Source	Share
Biomass	4.9%
Hydro	2.5%
Wind	2.2%
Solar*	0.6%
Geothermal energy	0.2%

Share of renewable energy consumption

RENEWABLE ENERGY SOURCES

Energy is an inevitable requirement where we want development to take place. All these power generation techniques can be described as renewable since they are not depleting any resources to create the energy. Renewable sources of energy are the ones which can be generated continuously in nature and are inexhaustible.

Biomass

Biomass or bio-energy, the energy from organic matter for thousands of years, ever since people started burning wood to cook food. Wood is still the largest biomass energy resource even today. Other sources of biomass can be used including plants, residues from agriculture or forestry and the organic components. Plants and animal matters are used for production of fibers chemicals or heat. The net emission of carbon dioxide will be zero as long as plants continue to be replenished for biomass energy purposes. Burning of plant or animal matters causes' air and water pollution. The burning of dung destroys essential nitrogen and phosphorus. Therefore, it is more useful to convert the biomass into biogas or bio fuels.

Biogas

Biogas is a mixture of methane, carbon dioxide, hydrogen and hydrogen sulphite, the major constituents being methane. Biogas is produced by anaerobic degradation of animal and plant wastes in the presence of water. Anaerobic degradation is to break down the organic matter by bacteria in the absence of oxygen. It is a non-polluting, clean and low cost fuel which is very useful for rural areas. Biogas plants used in the country are of two types; fixed dome biogas plant and floating drum biogas plant.

Tidal Energy

Tidal energy is not a very popular energy source, but has immense potential of becoming one in the near future. Tidal energy can be generated in two ways, tidal stream generators or by barrage generation. The power created through tidal generators is generally environmental friendly and causes less impact on established ecosystems. It is similar to the wind energy. Tidal energy is the only form of energy that derives directly from the motions of the Earth-Moon system. The tidal forces produced by the Moon-Sun in combination with Earth's rotation are responsible for the tides.

Wind Energy

Wind energy is a conversion of wind energy by wind turbines into a useful form, such as electricity or mechanical energy. Wind farms are installed on agricultural land or grazing areas, have one of the lowest environmental impacts of all energy sources. The principal application of wind power today is the generation of electricity, historically; it has been used directly to propel sailing ships or converted into mechanical energy for pumping water or grinding grains.

Geothermal Energy

Geothermal energy is the heat from Earth. It's clean and sustainable. Resources of geothermal energy range from the shallow ground to hot water and hot rock found a few miles beneath the Earth's surface and down even deeper to the extremely high temperatures of molten rock called magma. The steam or hot water comes out of the cracks in the Earth and when it doesn't find any way to come out, holes are drilled with pipes in it to gush the hot water out due to high pressure which turn the turbines of a generator to produce electricity.

Radiant Energy

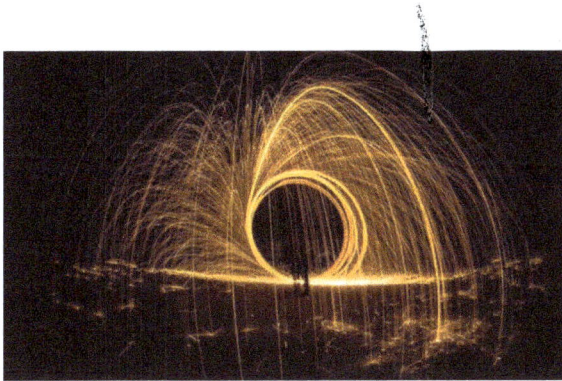

99% of the cost of normal electricity can be saved by the use of radiant energy. It performs the same functions, but doesn't possess behavior similar to electricity. Nikola Tesla's magnifying transmitter, T Henry Moray's radiant energy device, Edwin Gray's EMA motor & Paul Baumann's Testatika machine all ran on radiant energy. Nikola Tesla built one of the earliest wireless telephones to be based on radiant energy.

Hydro Electricity

This is the most widely used form of renewable energy. The gravitation force of falling water is the key point in hydroelectricity generation. Small scale hydro or micro-hydro power has been an increasingly popular alternative energy source, especially in remote areas where other power sources are not viable. The hydro power sites has a few major environmental problems like water logging and siltation, Causes loss to biodiversity of fish population and other aquatic animals. It also displaces local people and creates problems of rehabilitation and related socio-economic problems.

Compressed Natural Gas

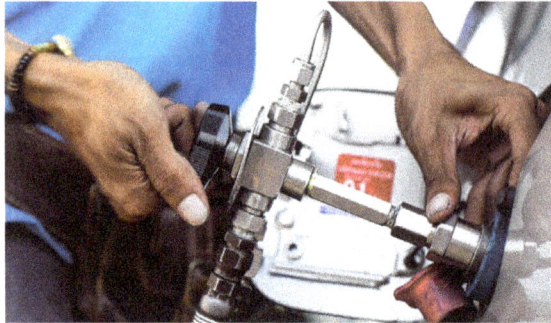

Compressed Natural Gas (CNG) is a substitute for gasoline, diesel or propane fuel. It is cleaner and safer to use as it diffuses easily into the surroundings if leaked. However, burning it does release a few greenhouse gases in the air. CNG is used in traditional gasoline internal combustion engine cars that have been converted into bi-fuel vehicles.

Solar Energy

The sun offers an ideal energy source, unlimited in supply, expensive, which does not add to the earth's total heat burden and does not produce air and water pollutants. Solar installations in recent years have also largely begun to expand into residential areas with government offering incentive programs to make "green" energy a more economically viable option.

Nuclear Energy

Proponents of nuclear energy contend that nuclear power is a sustainable energy source that reduces carbon emissions and increases energy security by decreasing dependence on foreign oil. Nuclear fission is used to extract energy from atomic nuclei via controlled nuclear reactions. Utility scale reactors are use to produce steam which is then converted into mechanical work for the purpose of generating electricity or propulsion.

ECONOMICS OF RENEWABLE ENERGY

Renewable energy (RE) technologies' market is on the rise, and the world is witnessing a new energy transition with many factors and drivers pushing it.

The histories of energy transitions, development of economies and industrial civilizations, all go hand in hand. Going back in time, people only needed to cover their basic needs, such as food, which -at the very beginning- was met by using firewood for cooking and heating. Further in time, people started practicing agriculture in the first formed human communities, essentially depending on the sun for that practice, in combination with biomass.

As economies evolved and developed into complex forms, firewood and other biomass were no lonager able to meet the increasing demand in energy.So people started turning into hydropower, then to coal during the 19th century, oil and natural gas in the 20th, in addition to nuclear that was introduced in mid-20th century as well.

Therefore, it is apprehendable that each critical change in the economic system -along history- was always accompanied with a major energy transition -and vice versa-, shifting from one major energy source to another. Currently, while fossil fuels (coal, oil and natural gas) are the dominant energy sources, the transition is already taking place from these sources into renewables (solar, wind, hydro, etc.).

Though, the 21st century energy transition is going underway, not mainly because of change in human needs, but due to other factors as well:

1. Concerns about environmental impacts (degradation, greenhouse gas emissions GHG, climate change, etc.).

2. The ongoing depletion of current energy sources, as they are limited and on the decline (millions of years to form, decades or less to consume).

3. The continuous price and technological change of different energy sources and their technologies.

Considering the added costs to mitigate, adapt to or fight the environmental side effects of using fossil fuels, renewables might be the only option that people/societies/governments have to adopt, in order to reform the current economic system —at least in the energy sector- into a new one.

Challenges to Consider

Assuming that renewable energy sources will actually be able to take hold in the near future, then a few questions need to be argued and discussed beforehand: What renewable energy sources are available? How to determine an optimal renewable energy mix? How will optimum mixtures of renewable-energy sources differ based on location? How to determine and calculate the direct and external costs of renewable energy sources? How will the existing achievements of the renewable-energy sector affect the way energy is processed in current economy? What kind of changes in sectors as engineering, economy and policy would be needed to adapt to renewable energy sources?

Scale is also an important issue. This is due to the fact that fossil-fuel technologies have been developed, improved and manufactured on an increasing scale for a century. This is not yet the case for renewables.

Economically, projections of energy sources' prices and their technologies are vital for forecasting the economic options of the energy supply, also with few critical questions in mind: Should the choice of a technology be based on its current market price or because of its potential future cost reductions? Which technologies offer the most effective outcomes for specific applications? If the current technology is too expensive, should governmental subsidies help to achieve cost reduction for economic viability or is it better to wait for market forces –Smith's invisible hands- to do the job?

Rationale for Renewables

Reasons which have contributed to the acceleration of both public and private investment in renewable energy:

1. The growing demand for energy, which consequently requires a certain economic development.

2. The fact that fossil fuels are finite, and negatively affecting the climate and polluting the air.

3. The current critical environmental and climatic conditions, which drive the need to redirect energy technologies into more diverse, environmentally sustainable supply sources.

4. The need to ensure future energy security.

5. Mostly for developing countries in particular: Rapid urbanization, economic growth, uprising demographic trends and severe climate change conditions.

Utilizing renewables would help to avoid these problems, create new job opportunities and reduce the drain on hard currency for poorer countries. Because conventional fuels have received long-term subsidies in the past, it is vital that governments support the development of renewables in the form of financial incentives that can create a level playing field.

The future of the renewables industry depends on finance, risk-return profiles, business models, lifetime's investment and a sum of other economic, policy and social factors. Many new sources of finance are possible such as insurance funds, pension funds and sovereign wealth funds along with new mechanisms for financial risk mitigation. Many new business models are also possible for local energy services, utility services, transport, community and cooperative ownership, and rural energy services.

Global Investements in Renewables

In 2011, the global investment in renewable power and fuels increased to a new record. Significantly, developing economies made up 35% of this total investment. In addition, the whole period 2004-2017 has witnessed a remarkable increase in investments in renewables, either in different sectors, or for different technologies, in different countries with different economic systems, as illustrated in the following figures.

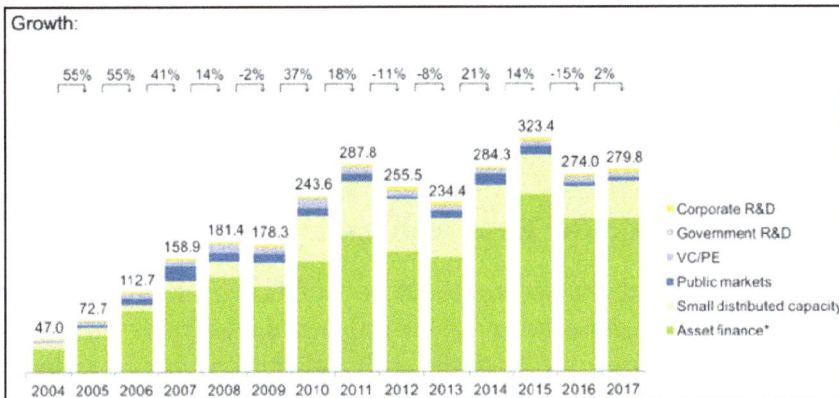

New Global Investments ($Bn) in Renewables by Asset Class 2004-2017.

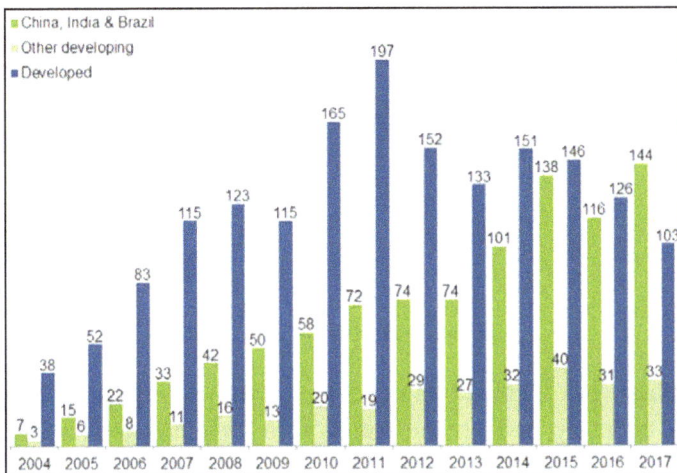

New Global Investments in Renewables by the Type of Economy 2004-2017.

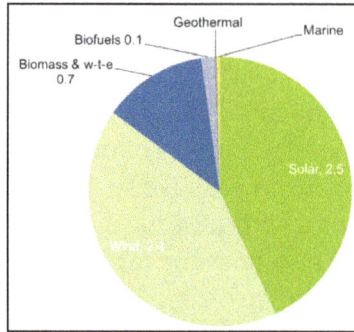

Public Markets Investment ($Bn) in RE by Sector in 2017.

Renewable energy technologies (RETs) continue to face a number of barriers. However, the major challenge is mainly economic, as the issue of renewable energy technologies' costs is vital and central for the prediction of how rapidly the current energy transition will be taking place. The costs include: infrastructure investment, day-to-day operations, market costs of supply and the environmental costs of the different energy sources.

Therefore, the debate remains mainly focused on the economic and financial perspectives, particularly on the cost-effectiveness of renewable energy technologies, and the possible various economic incentives to promote renewables globally in terms of: regulatory design and affordability.

Economic Rationale for Renewables

While by 2014 the world was getting about 80% of its electricity supplies from fossil fuels, that percentage has gone down 3.5-4% only within 3-4 years. In 2017/18 fossil fuels contributed approximately 76.5% to the global electricity supply, reflecting the rise in the global renewables' market.

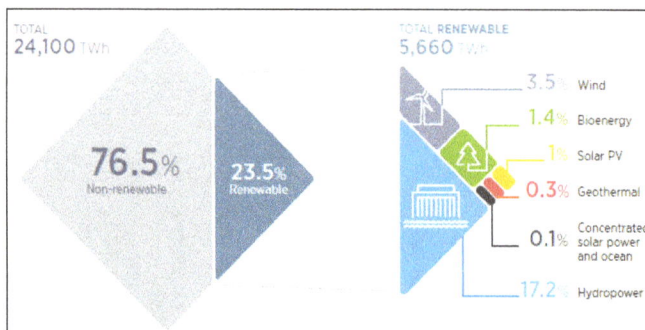

Global Electricity Generation.PNG.

The cost advantage that fossil fuels used to have over renewable energy sources has been decreasing recently, with some renewable technologies (Solar PV, wind, hydropower) already competing fossil fuels directly on the financial frontier. Furthermore, renewables' costs are expected to decline even further, and those of fossil fuels will incline. The following two figures show that -while on one hand- the oil prices are on the rise during the 2000s, on the other hand, investments in renewables are on the rise during the same period, thus reflecting its competitiveness against oil in recent years.

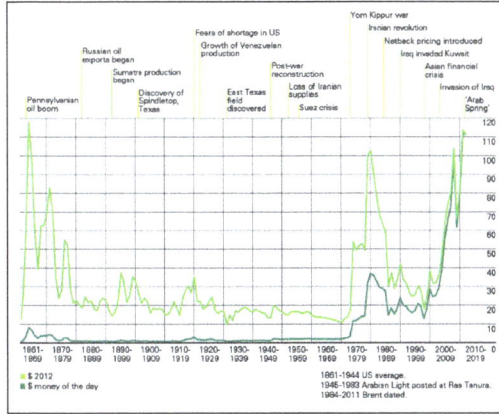

The Change in Crude Oil Prices (USD per Barrel) Since 1861 with Accordance to Global Major Events.

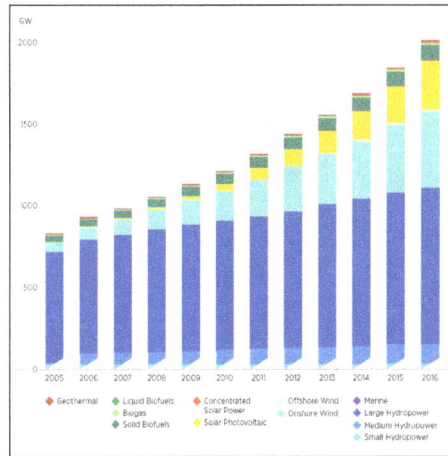

Trends in RE Installed Capacity by Technology.

The renewables' market development during the past 10-15 years had few moving factors, which can be summarized as follows:

1. One outcome of the Kyoto Protocol, entering into force in early 2005, was the exponential growth of global investment in renewables.

2. Rapid growth in energy demand for emerging economies, such as the cases of China & India, which are driven by transforming their energy industries.

3. Uprising competition for energy sources.

4. Inclining geopolitical tension.

5. Energy security concerns.

6. Increasing prices of oil and gas.

7. Technological developments in the renewables' sector, and the emergence of more technology applications, especially generation of solar PV and wind power, which actually alone makes renewables more competitive, even without investment support.

8. The need to commit to a long-term sustainable energy targets has further improved the climate for investments in renewables.

9. Positive support of policy and law-makers in various countries, promoting scarcity of fossil fuels, their on-the-rise prices and climate challenges, which require adopting different energy approaches.

10. Intensive research efforts, leading to improved system solutions with much higher efficiencies and lower production and operation costs.

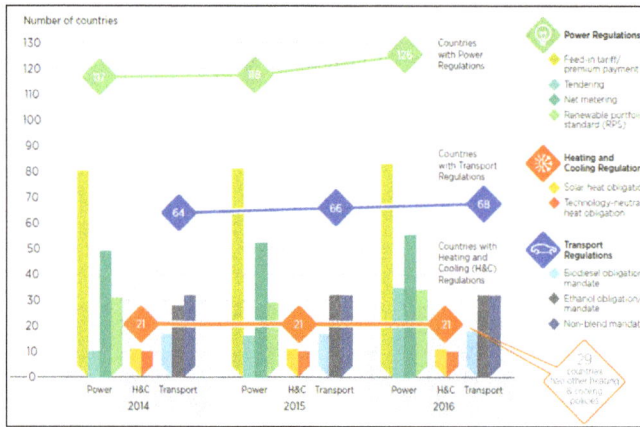

RE Costs Declining, Policy & Regulations in Different Sectors Inclining.

Market Situation

Costs of Renewables

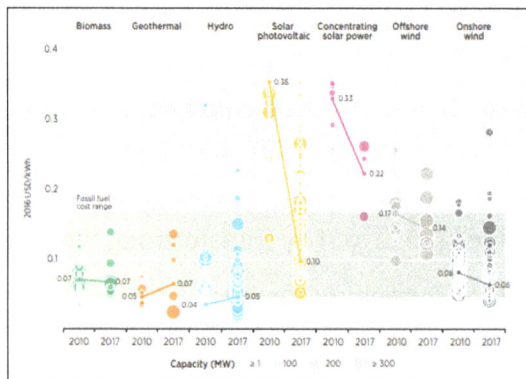

Global Levelized Cost of Energy (LCOE) from Utility-Scale Renewable
Power Generation Technologies.

According to the most recent reports on renewable energy technologies, from IRENA, REN21 and IEA, electricity costs from almost all the renewable projects that were commissioned in 2017, have continued to decline. Projects of bioenergy power, hydropower, geothermal and onshore wind, which were commissioned in that year, have widely fallen into the generation costs' range of fossil-generated electricity, and furthermore, some of these projects have actually undercut those of fossil fuels-based ones.

The most common methodology for comparing different energy sources, is to calculate the Levelized Cost Of Energy (LCOE). LCOE measures lifetime costs, including building and operation of a power plant, divided by lifetime energy production/output.

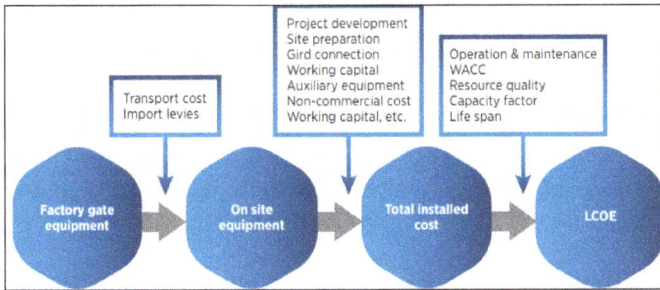

Cost metric analysis for the calculation of LCOE.

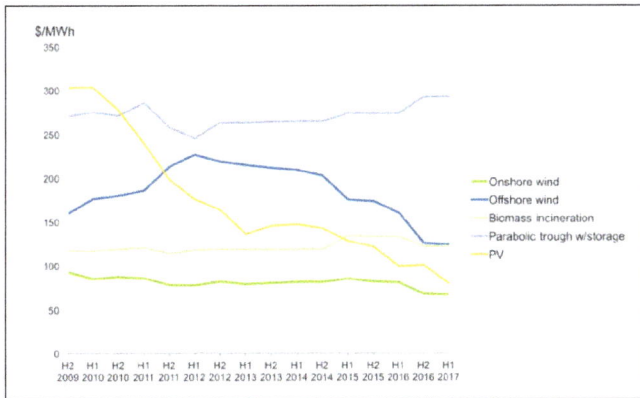

LCOE of Electricity by Renewables ($ per MWH) 2009-2017.

As shown in the figure above, Global weighted LCOE of utility-scale solar PV has witnessed a remarkable drop (approximately 27%) since 2010, reaching USD 0.10/kWh for the new commissioned projects in 2017. Under the right conditions, it will potentially decline to USD 0.03/kWh from 2018 onward.

Onshore wind is already one of the most competitive sources for generation capacity. Recent auctions in Brazil, Canada, Germany, India, Mexico and Morocco have resulted in LCOE as low as USD 0.03/kWh.

On the other hand, many auctions predict that by 2020, both Concentrated Solar Power (CSP) & offshore wind would have the potential to provide electricity with LCOE within the range of USD 0.06 - 0.10/kWh.

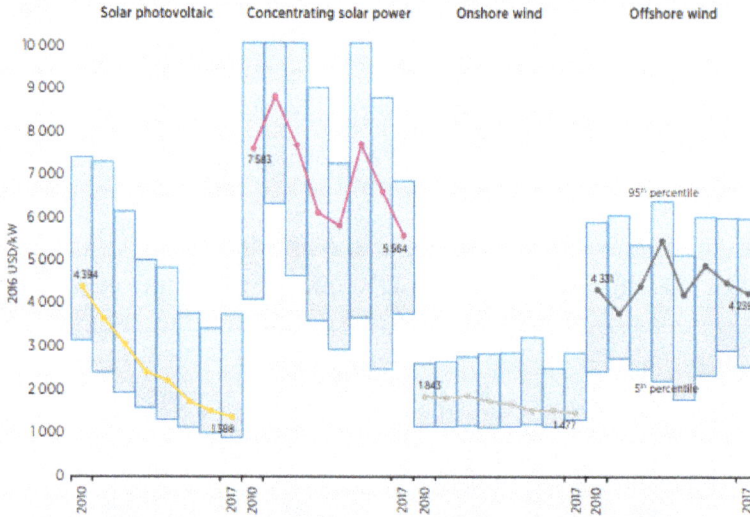

Global weighted average total installed costs and project percentage
ranges for CSP, Solar PV, Onshore & Offshore wind 2010-2017

The varying fall ranges in LCOE for solar and wind power in particular have been mainly driven by the reduction in total installment costs, which is affected by three main
forces:

- Technology improvements.

- Competitive procurement and the rise of patents and innovators in the sector.

- The consequent emergence of a large base of experienced medium-to-large
  project developers, who are actively seeking new markets globally.

Table: Costs' Fall Indicators for Solar & Wind Technologies.

	% Drop in Installed Costs	Period	% Drop in LCOE
Solar PV Modules	68	2010-2017	73
Concentrated Solar Thermal Projects	27	2010-2017	33
Onshore Wind Projects	20	2010-2017	22
Offshore Wind Projects	2	2010-2017	13

Based on current installed projects and auction data, in combination with mass production increase and specific investment costs, electricity from renewables -sooner
rather than later- will be cheaper than that from fossil fuels. All the renewable power
generation technologies are expected to fall within the fossil fuel cost range, with the
majority having the potential to undercut it. This will significantly lower the LCOE of
all technologies, eventually leading to a market potential increase and development for
renewables.

Costs of Fossil Fuels

Costs relative to fossil fuels are also important particularly because:

- Fossil-fuel energy does not reflect its full social costs.

  ○ Climate change has been described as the "biggest market failure in history" because the environmental costs associated with carbon emissions are not included in market prices.

  ○ Furthermore, fossil fuels are subsidized for about US$300 billion per year. Removing theses subsidies and incorporating externalities into fossil fuel costs would dramatically change relative costs.

  ○ Also the external costs which are related to the use of fossil fuels, stemming from different causes: pollution and environmental degradation as a consequence of extraction of resources, indoor and outdoor air pollution, resulting from direct fuel combustion, as well as non-combustion emissions (e.g. industrial processes).

  ○ Further side effects, which also could add to the externalities, as fossil fuels produce: Sulphur oxide SO_2, mono-nitrogen oxides NO_x, particular emissions $PM_{2.5}$, ammonia NH_3 and volatile organic compounds VOCs, which can cause: adverse human health effects, reducing agricultural yields, damaging forests, buildings, and infrastructure.

  ○ Currently, the climate change and air pollution's external effects -alone- are approximately in the range of 2.2-5.9 trillion USD per year, while the all-in-all cost of the global energy supply is around 5 trillion USD per year. The externalities of air pollution, caused by fossil fuels in Europe alone, were recorded ranging between 330 billion-940 billion USD in 2010.

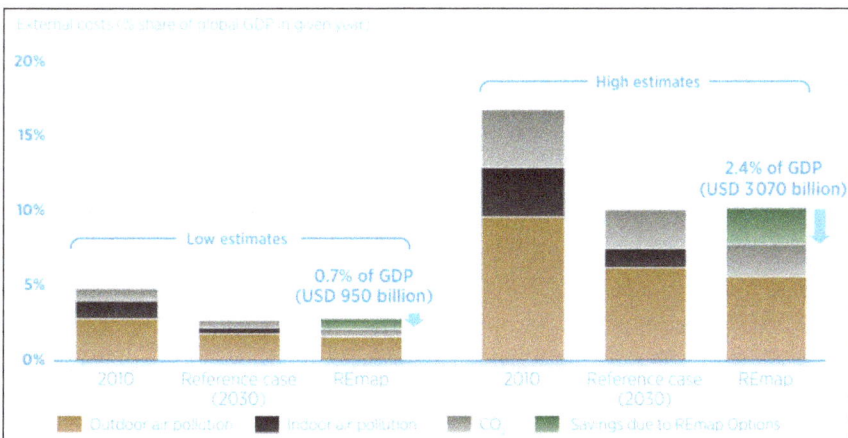

External Costs as a Share of GDP & Possible Reduced Externalities
with IRENA's Renewable Energy Map Tool's Options.

- It is more expensive to deliver non-renewable energy in some places than others.

    ○ For example, rural or remote communities in developing countries are often not connected to the grid, resulting in "off-grid" energy production - particularly solar power - being more competitive than extending the grid.

- Fossil Fuels are fast depleting and scarcer than RE.

    ○ Fossil fuels are finite, as upon being consumed long enough, global resources will eventually run out.

    ○ For a proper estimate of how long can current fossil fuel reserves be consumed for, the following figure, has been plotted by dividing the quantity of known fuel reserves by the current rate of production.

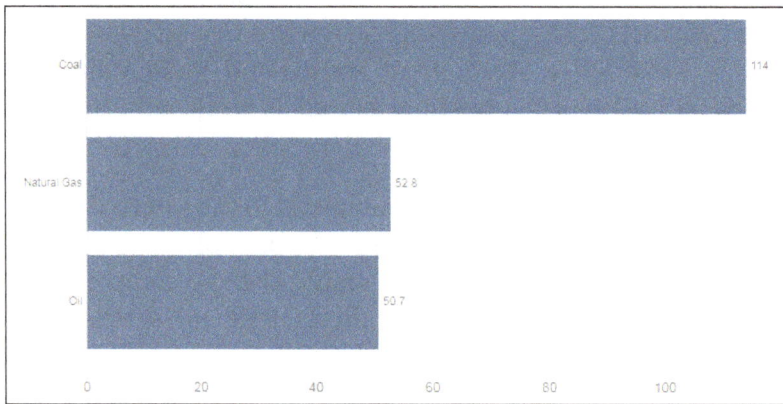

Years of Fossil Fuel Reserves Left.

Market Competition - Renewables vs. Fossil Fuels

As the markets develop, the costs normally do as well, as both developments go hand in hand. The previously mentioned factors push the market to increase its renewables' volume, leading to economies of scale. On one hand, this reduces the price and later the actual costs of the technology, while on the other hand, reduced prices increase market volumes, again producing economies of scale, eventually resulting in a feedback loop, that either way paves the path for renewables.

The continuous pressure on market prices and its margins is rapidly forcing the market to change, as renewables' costs have considerably declined and are still on the decline. Their costs are expected to go down even further over the coming few years. Furthermore, adding to renewables' economic evolution, both public commitments and the maturing technologies, investments in renewables have rapidly increased turning the renewables industry to a very competitive sector against other energy resources. However, the competition is not only limited within the energy or power sector itself, but different renewables are even starting to compete against each other within the renewables' sector itself.

Table: Cost Development of RE different technologies.

Technology	Market & Costs' Development	Why?	Future Projections
Solar PV	• Rapidly declining annually • Declined by 58% between 2010-2015 • Modules are 80% cheaper than they were in 2009 • Cost of generated electricity dropped to 3/4 and continue to decline 2010-2017	• PV modules' technology & manufacturing improvements • Rapid deployment	• Trend is likely to continue • Another 57% drop by 2025
Wind Power	• Has been the most competitive renewable technology against fossil fuels technologies since 2015 • 50% price drop 2010-2017 • Onshore wind electricity costs have dropped by approximately 25% since 2010	• Their prices have already dropped since 1990s • Remained steady along the past decade	• Further technological development and price drop • Further drop in the overall generation costs, as the average capacity factor grew, turbines are more efficient, generating more electricity per turbine
Concentrated Solar Power (CSP) & Solar Thermal Energy (STE)	• Decreased in costs • Parallelly moved into new market sector • Specific generation costs per MWh significantly fell		
Hydropower	• The overall market volume in the past decade was greater than earlier decades	• Very mature as it is the oldest RE technology, since 1868	• A little chance to further cost reduction
Biomass Power	• Competitive power generation option wherever low-cost agricultural or forestry waste is available		• New technologies are emerging, hence there is potential for cost reduction

With costs of renewables are continuing to fall, drastically in solar PV, followed by wind and concentrated solar power closely behind, the global installed capacity has exponentially grown. A world record amount of recently installed renewables' (especially solar PV, wind, CSP & hydro) capacity has been added in the past few years. Thus adding up to almost two thirds of the all new generating capacity installed globally in 2016.

Renewables' power capacity investments have by far surpassed those of fossil fuels in the year 2017. The renewable energy market has been catalyzed by increasing innovation, competition and policy support. Hence, radical technological advances and sharp cost reduction in renewables' sector have been achieved, pushing renewables to outpace any other technology source.

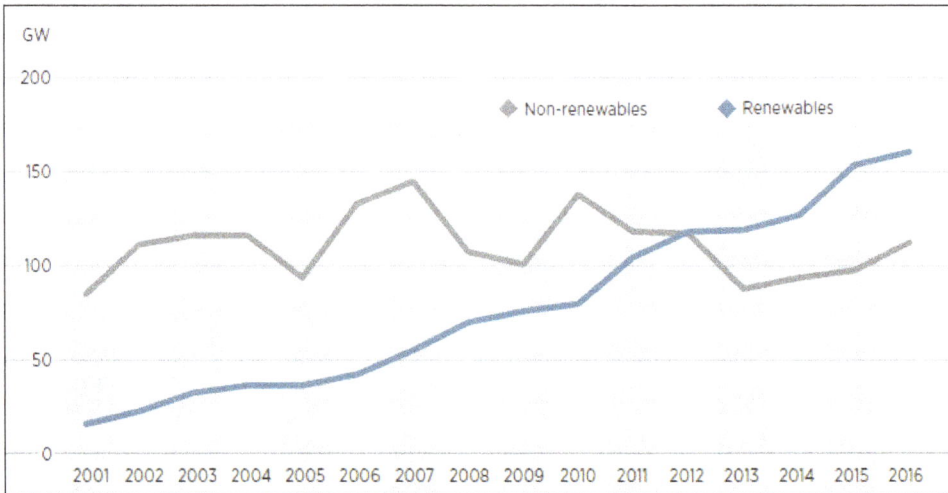

Renewables vs. Non Renewables in Global Capacity Additions 2001-2016

Economically Accelerating the Energy Transition to Renewables

Though renewables' market is inclining, and most probably will do so for the coming decades, most of the recent reports suggest that it would still not be enough to meet the global goals by 2030. Therefore, the following presents some strategies that can push and encourage investment in the sector.

Expansion of Renewable Energy use:

1. The basic economics of renewable energy need to be artificially altered, either by increasing the cost of fossil fuel-based energy (e.g. through taxes, removing subsidies or equivalent mechanisms), or by reducing the costs of renewable energy (e.g. subsidies), or by boosting the returns to renewable energies (e.g. through paying a premium for this form of energy). Removal or gradually reducing governmental fossil fuels subsidies is being carried out in some cases (e.g. Egypt in the past 2-3 years).

2. Developing countries should not necessarily be required to meet these costs. This is particularly so where the development of renewable energy capacity may place countries at a competitive disadvantage and/or these countries bear no responsibility for climate change. The costs should be met by countries that do bear these responsibilities.

3. Declining renewables' costs, which is also already taking place.

4. Implementing new renewables' financial policies.

The following points provide a set of requirements and recommendations for a successful and more efficient cost reduction policies for renewables:

1. Encouraging domestic manufacturing of renewables' equipment: the example of the Chinese case would be the best to illustrate this point, since the Chinese low-cost equipment have achieved a lot for the promotion of affordable renewable projects around Asia.

2. Reducing institutional barriers: experience has shown that institutional dysfunction always leads to delays, consequently having a major impact on the economic value of the projects in hands.

3. Grounding renewables in the economic analysis and applying market principles.

4. Enhancing transmission grids and supporting transmission integration.

TECHNOLOGICAL ADVANCES IN RENEWABLE ENERGY

Not only is the renewable energy industry expanding, but also seeing a lot of change within itself and the difference a few years can make in the ongoing transition is remarkable. Less than a decade ago, uncertainty about high generation costs still overshadowed the rise of solar and wind power. For those who were paying attention, it was inevitable that technological improvements, economies of scale, increased competition in supply chains and the right political conditions had begun a continuous process, driving down the cost from renewables.

As of today, the competitiveness of renewable energy alternatives has become increasingly clear to everyone. Yet hard work continues as governments, industry, and investors plan for the next phase of the energy transition. This involves proactive discussions to create new policies, regulations, market structures, and industrial strategies, in particular by supporting the stable integration of the highest levels of renewable electricity generation. We also need strategies to reduce end-use CO_2 emissions from transport and industry which brings the role of electricity storage to the center of the stage. As of today, the competitiveness of renewable energy alternatives has become increasingly clear to everyone.

Energy Storage

Advanced energy storage systems provide a wide array of technological approaches to managing our power supply in order to create a more resilient energy infrastructure and bring cost savings to utilities and consumers. Important is that it brings the flexibility that future electricity systems need to accommodate the fluctuating availability of solar and wind energy. For the longer-term, as countries strive to significantly reduce emissions from power generation, the importance of storage will only increase.

Battery storage technology is multifaceted – a complex technology. Its economics can be shaped by such things as customer type, location, grid needs, regulations, rate structure, and nature of the application. It is also uniquely flexible in its ability to stack value streams and change its dispatch to serve different needs over the course of a year or even an hour. These value streams are growing both in value and in market scale and will serve a remarkable role in the future.

Although lithium-ion batteries have received the most attention so far, other types are becoming more and more cost-effective. Battery storage can be deployed both on the grid and at an individual consumer's home or business and is poised to play a decisive role in the transition to a sustainable energy future. It is estimated to grow at least seventeen times by 2030. Huge investments have already been made, as in September 2018, a major announcement was made by the World Bank Group (WBG) who committed $1 billion for a new global program to accelerate investments in battery storage for energy systems in developing and middle-income countries. This investment is intended to increase developing countries' use of wind and solar power, and improve grid reliability, stability, and power quality while reducing carbon emissions.

Digitalization

Digitalization enables the integration of renewable energy. Artificial intelligence and analytics are helping facility owners and operators optimize their renewable energy output. Nascent Blockchain technology is helping to get renewables on the grid and offers a way for untrusted parties the change to reach a common digital history. This is so digital assets and transactions cannot be easily faked or duplicated despite not having a trusted intermediary. Simply put, the blockchain technology enables excess output from wind and solar to be discharged as needed into a networked pool of home battery storage systems in real time.

Solar Technology

There are two main types of solar technology: photovoltaics (PV) and concentrated solar power (CSP). Solar PV technology captures sunlight to generate electric power, and CSP harnesses the heat of the sun and uses it to produce thermal energy that powers heaters or turbines. These two forms of solar energy enable a wide range of technical innovation.

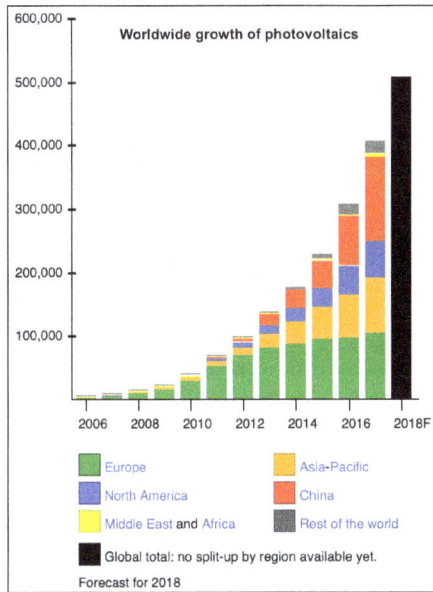

Approximate regional shares estimated from IEA.

1. Advances in solar panel efficiency: There has been an ongoing race in the solar industry in terms of solar cell efficiency which has been leading to the research of alternative solar cell types and there have been some major breakthroughs in the past two years, especially with Perovskite solar cells (silicon cells are mostly used today). There is also a new tech concept revealed that captures and utilizes the waste heat that is usually emitted by solar panels. While this typically released and non-harnessed thermal energy is a setback and an opportunity for improvement in solar technology, this innovation could help reduce solar costs even more, which could double the efficiency of solar cells.

2. Solar skin design: A concept of aesthetic enhancement that allows solar panels to have a customized look that makes it possible for solar panels to match the appearance of a roof without interfering with panel efficiency or production.

3. Solar powered roads: These roadways have the ability to generate clean energy through modular solar panels, and they also include LED bulbs that can light roads at night and have a thermal heating capacity that can melt snow during winter weather. The panels are also used in bike lanes and the most famous one is located in Krommenie, Netherlands.

4. Wearable solar: This new textile concept makes it possible tiny solar panels can now be stitched into the fabric of clothing, suitable for home products like window curtains etcetera.

5. Solar tracking mounts: Trackers allow solar panels to maximize electricity production by following the sun as it moves across the sky. PV tracking systems tilt and shift the angle of a solar array as the day goes by to best match the sun's position.

6. Solar water purifiers: The new product can access visible light and only requires a few minutes to produce reliable drinking water while prior purifier designs needed to harness UV rays and required hours of sun exposure to fully purify water. This means remarkable improvements in efficiency compared to past technology.

7. Solar thermal fuel (STF): Is a material that is capable of absorbing photons (light) and storing their energy as a charge and then releasing it when prompted. This is an alternative storage solution for solar. We have seen this technology in recent years implemented for windows and other surfaces that are exposed to sunlight.

BARRIERS TO RENEWABLE ENERGY TECHNOLOGIES

Capital Costs

The most obvious and widely publicized barrier to renewable energy is cost—specifically, capital costs, or the upfront expense of building and installing solar and wind farms. Like most renewables, solar and wind are exceedingly cheap to operate—their "fuel" is free, and maintenance is minimal—so the bulk of the expense comes from building the technology.

The average cost in 2017 to install solar systems ranged from a little over $2,000 per kilowatt (kilowatts are a measure of power capacity) for large-scale systems to almost

$3,700 for residential systems. A new natural gas plant might have costs around $1,000/kW. Wind comes in around $1,200 to $1,700/kw.

Renewables are cheap to operate, but can be expensive to build.

Higher construction costs might make financial institutions more likely to perceive renewables as risky, lending money at higher rates and making it harder for utilities or developers to justify the investment. For natural gas and other fossil fuel power plants, the cost of fuel may be passed onto the consumer, lowering the risk associated with the initial investment (though increasing the risk of erratic electric bills).

Selecting an appropriate site for renewables can be challenging.

However, if costs over the lifespan of energy projects are taken into account, wind and utility-scale solar can be the least expensive energy generating sources, according to asset management company Lazard. As of 2017, the cost (before tax credits that would further drop the costs) of wind power was $30-60 per megawatt-hour (a measure of energy), and large-scale solar cost $43-53/MWh. For comparison: energy from the most efficient type of natural gas plants cost $42-78/MWh; coal power cost at least $60/MWh.

Even more encouragingly, renewable energy capital costs have fallen dramatically since the early 2000s, and will likely continue to do so. For example: between 2006 and 2016, the average value of photovoltaic modules themselves plummeted from $3.50/watt $0.72/watt—an 80 percent decrease in only 10 years.

Siting and Transmission

Nuclear power, coal, and natural gas are all highly centralized sources of power, meaning they rely on relatively few high output power plants. Wind and solar, on the other hand, offer a decentralized model, in which smaller generating stations, spread across a large area, work together to provide power.

Decentralization offers a few key advantages (including, importantly, grid resilience), but it also presents barriers: siting and transmission.

Siting is the need to locate things like wind turbines and solar farms on pieces of land. Doing so requires negotiations, contracts, permits, and community relations, all of which can increase costs and delay or kill projects.

Transmission refers to the power lines and infrastructure needed to move electricity from where it's generated to where it's consumed. Because wind and solar are relative newcomers, most of what exists today was built to serve large fossil fuel and nuclear power plants.

But wind and solar farms aren't all sited near old nuclear or fossil fuel power plants (in fact, some areas with fewer older power plants, such as the Great Plains and Southwest, offer some of the country's best renewable potential). To adequately take advantage of these resources, new transmission infrastructure is needed—and transmission costs money, and needs to be sited. Both the financing and the siting can be significant barriers for developers and customers, even when they're eager for more renewables—though, again, clean energy momentum is making this calculation easier.

Market Entry

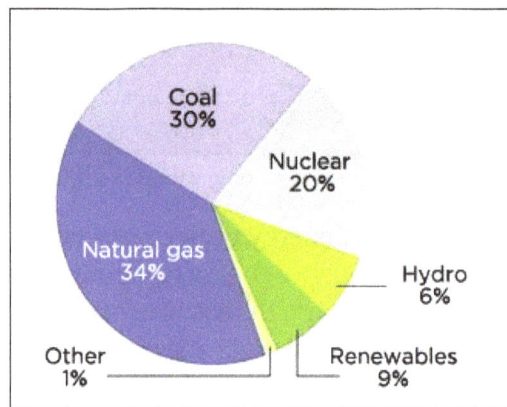

Renewables face stiff competition
from more established, higher-carbon sectors.

For most of the last century US electricity was dominated by certain major players, including coal, nuclear, and, most recently, natural gas. Utilities across the country have invested heavily in these technologies, which are very mature and well understood, and which hold enormous market power.

This situation—the well-established nature of existing technologies—presents a formidable barrier for renewable energy. Solar, wind, and other renewable resources need to compete with wealthier industries that benefit from existing infrastructure, expertise, and policy. It's a difficult market to enter.

New energy technologies—startups—face even larger barriers. They compete with major market players like coal and gas, and with proven, low-cost solar and wind technologies. To prove their worth, they must demonstrate scale: most investors want large quantities of energy, ideally at times when wind and solar aren't available. That's difficult to accomplish, and a major reason why new technologies suffer high rates of failure.

Increased government investment in clean energy—in the form of subsidies, loan assistance, and research and development—would help.

Climate action opponents like former EPA administrator Scott
Pruitt have long been propped up by industry money.

Unequal Playing Field

Oil Change International estimates that the United States spends $37.5 billion on subsidies for fossil fuels every year. Through direct subsidies, tax breaks, and other incentives and loopholes, US taxpayers help fund the industry's research and development, mining, drilling, and electricity generation. While subsidies have likely increased domestic production, they've also diverted capital from more productive activities (such as energy efficiency) and constrained the growth of renewable energy (solar and wind enjoy fewer subsidies and, generally, receive much less preferential political treatment).

For decades, the fossil fuel industry has used its influence to spread false or misleading information about climate change—a strong motivation for choosing low-carbon energy sources like wind or solar (in addition to the economic reasons). Industry leaders knew about the risks of global warming as early as the 1970s, but recognized that dealing with global warming meant using fewer fossil fuels. They went on to finance—and continue to fund—climate disinformation campaigns, aimed at sewing doubt about climate change and renewable energy.

Their efforts were successful. Despite widespread scientific consensus, climate action is now a partisan issue in the US congress, complicating efforts to move from fossil fuels to clean energy.

The disconnect between science and policy means that the price we pay for coal and gas isn'trepresentative of the true cost of fossil fuels (ie, it doesn't reflect the enormous costs of global warming and other externalities). This in turn means that renewables aren't entering an equal playing field: they're competing with industries that we subsidize both directly (via government incentives) and indirectly (by not punishing polluters).

Emission fees or caps on total pollution, potentially with tradable emission permits, are examples of ways we could use to help remove this barrier.

Reliability Misconceptions

When done correctly, reliability isn't a concern with wind and solar—it's actually a strength.

Renewable energy opponents love to highlight the variability of the sun and wind as a way of bolstering support for coal, gas, and nuclear plants, which can more easily operate on-demand or provide "baseload" (continuous) power. The argument is used to undermine large investments in renewable energy, presenting a rhetorical barrier to higher rates of wind and solar adoption.

But reality is much more favorable for clean energy. Solar and wind are highly predictable, and when spread across a large enough geographic area—and paired with complementary generation sources—become highly reliable. Modern grid technologies like advanced batteries, real-time pricing, and smart appliances can also help solar and wind be essential elements of a well-performing grid.

Many utilities, though, still don't consider the full value of wind, solar, and other renewable sources. Energy planners often consider narrow cost parameters, and miss the big-picture, long-term opportunities that renewables offer. Increased awareness—and a willingness to move beyond the reliability myth—is sorely needed.

References

- Renewable-energy-clean-facts, stories: nrdc.org, Retrieved 31 March, 2019

- Public-benefits-of-renewable-power, renewable-energy, clean-energy: ucsusa.org, Retrieved 14 July, 2019

- Advantages-disadvantages-renewable-energy: brandongaille.com, Retrieved 17 May, 2019

- Top-10-renewable-energy-sources: atomberg.com, Retrieved 19 April, 2019

- The-economics-of-renewable-energy: energypedia.info, Retrieved 5 February, 2019

- Technological-advances-in-renewable-energy: theenergybit.com, Retrieved 26 July, 2019

- Barriers-to-renewable-energy, renewable-energy, clean-energy: ucsusa.org, Retrieved 21 May, 2019

2

Solar Energy and Photovoltaic Systems

The radiation from the sun that is capable of producing heat by causing chemical reactions is referred to as solar energy. Some of the important aspects of solar energy are solar power and solar thermal energy. This chapter discusses in detail these aspects related to solar energy as well as solar panels and photovoltaic systems.

Solar energy is the radiation from the Sun capable of producing heat, causing chemical reactions, or generating electricity. The total amount of solar energy incident on Earth is vastly in excess of the world's current and anticipated energy requirements. If suitably harnessed, this highly diffused source has the potential to satisfy all future energy needs. In the 21st century solar energy is expected to become increasingly attractive as a renewable energy source because of its inexhaustible supply and its nonpolluting character, in stark contrast to the finite fossil fuels coal, petroleum, and natural gas.

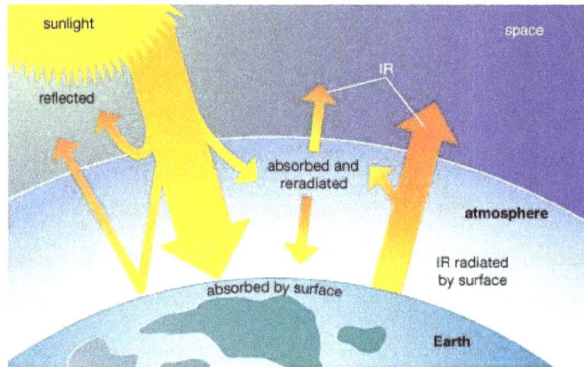

Solar energy: Reflection and absorption of solar energy. Although some incoming sunlight is reflected by Earth's atmosphere and surface, most is absorbed by the surface, which is warmed.

The Sun is an extremely powerful energy source, and sunlight is by far the largest source of energy received by Earth, but its intensity at Earth's surface is actually quite low. This is essentially because of the enormous radial spreading of radiation from the distant Sun. A relatively minor additional loss is due to Earth's atmosphere and clouds, which absorb or scatter as much as 54 percent of the incoming sunlight. The sunlight that reaches the ground consists of nearly 50 percent visible light, 45 percent infrared radiation, and smaller amounts of ultraviolet and other forms of electromagnetic radiation.

Uses of Solar Energy

The potential for solar energy is enormous, since about 200,000 times the world's total daily electric-generating capacity is received by Earth every day in the form of solar energy. Unfortunately, though solar energy itself is free, the high cost of its collection, conversion, and storage still limits its exploitation in many places. Solar radiation can be converted either into thermal energy (heat) or into electrical energy, though the former is easier to accomplish.

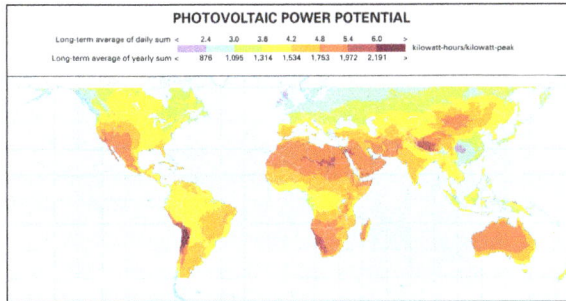

Solar energy potential:
Earth's photovoltaic power potential.

Thermal Energy

Among the most common devices used to capture solar energy and convert it to thermal energy are flat-plate collectors, which are used for solar heating applications. Because the intensity of solar radiation at Earth's surface is so low, these collectors must be large in area. Even in sunny parts of the world's temperate regions, for instance, a collector must have a surface area of about 40 square metres (430 square feet) to gather enough energy to serve the energy needs of one person.

The most widely used flat-plate collectors consist of a blackened metal plate, covered with one or two sheets of glass, that is heated by the sunlight falling on it. This heat is then transferred to air or water, called carrier fluids, that flow past the back of the plate. The heat may be used directly, or it may be transferred to another medium for storage. Flat-plate collectors are commonly used for solar water heaters and house heating. The storage of heat for use at night or on cloudy days is commonly accomplished by using insulated tanks to store the water heated during sunny periods. Such a system can supply a home with hot water drawn from the storage tank, or, with the warmed water flowing through tubes in floors and ceilings, it can provide space heating. Flat-plate collectors typically heat carrier fluids to temperatures ranging from 66 to 93 °C (150 to 200 °F). The efficiency of such collectors (i.e. the proportion of the energy received that they convert into usable energy) ranges from 20 to 80 percent, depending on the design of the collector.

Another method of thermal energy conversion is found in solar ponds, which are bodies of salt water designed to collect and store solar energy. The heat extracted from such ponds enables the production of chemicals, food, textiles, and other industrial products and can also be used to warm greenhouses, swimming pools, and livestock buildings. Solar ponds are sometimes used to produce electricity through the use of the organic Rankine cycle engine, a relatively efficient and economical means of solar energy conversion, which is especially useful in remote locations. Solar ponds are fairly expensive to install and maintain and are generally limited to warm rural areas.

Solar heating: A building roof with flat-plate collectors that
capture solar energy to heat air or water.

On a smaller scale, the Sun's energy can also be harnessed to cook food in specially designed solar ovens. Solar ovens typically concentrate sunlight from over a wide area to a central point, where a black-surfaced vessel converts the sunlight into heat. The ovens are typically portable and require no other fuel inputs.

A monk using a solar-powered cookstove in the Potala Palace, Lhasa, Tibet.

Electricity Generation

Solar radiation may be converted directly into electricity by solar cells (photovoltaic cells). In such cells, a small electric voltage is generated when light strikes the junction between a metal and a semiconductor (such as silicon) or the junction between two different semiconductors. The power generated by a single photovoltaic cell is typically only about two watts. By connecting large numbers of individual cells together, however, as in solar-panel arrays, hundreds or even thousands of kilowatts of electric power can be generated in a solar electric plant or in a large household array. The energy efficiency of most present-day photovoltaic cells is only about 15 to 20 percent, and, since the intensity of solar radiation is low to begin with, large and costly assemblies of such cells are required to produce even moderate amounts of power.

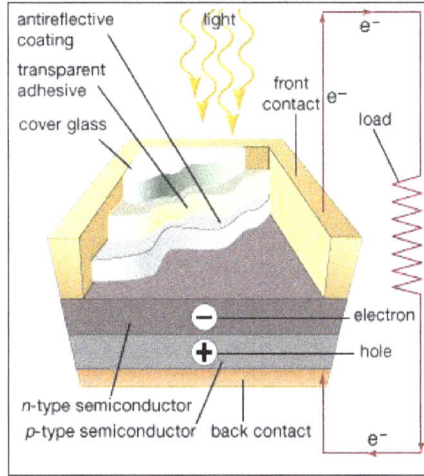

In the above figure when sunlight strikes a solar cell, an electron is freed by the photo-electric effect. The two dissimilar semiconductors possess a natural difference in electric potential (voltage), which causes the electrons to flow through the external circuit, supplying power to the load. The flow of electricity results from the characteristics of the semiconductors and is powered entirely by light striking the cell.

Small photovoltaic cells that operate on sunlight or artificial light have found major use in low-power applications—as power sources for calculators and watches, for example. Larger units have been used to provide power for water pumps and communications systems in remote areas and for weather and communications satellites. Classic crystalline silicon panels and emerging technologies using thin-film solar cells, including building-integrated photovoltaics, can be installed by homeowners and businesses on their rooftops to replace or augment the conventional electric supply.

Solar power: Single-family house with solar panels on the roof.

Concentrated solar power plants employ concentrating, or focusing, collectors to concentrate sunlight received from a wide area onto a small blackened receiver, thereby considerably increasing the light's intensity in order to produce high temperatures.

The arrays of carefully aligned mirrors or lenses can focus enough sunlight to heat a target to temperatures of 2,000 °C (3,600 °F) or more. This heat can then be used to operate a boiler, which in turn generates steam for a steam turbine electric generator power plant. For producing steam directly, the movable mirrors can be arranged so as to concentrate large amounts of solar radiation upon blackened pipes through which water is circulated and thereby heated.

Concentrated solar-power plant
Nevada Solar One, a concentrated solar-power plant.

Other Applications

Solar energy is also used on a small scale for purposes. In some countries, for instance, solar energy is used to produce salt from seawater by evaporation. Similarly, solar-powered desalination units transform salt water into drinking water by converting the Sun's energy to heat, directly or indirectly, to drive the desalination process.

Solar technology has also emerged for the clean and renewable production of hydrogen as an alternative energy source. Mimicking the process of photosynthesis, artificial leaves are silicon-based devices that use solar energy to split water into hydrogen and oxygen, leaving virtually no pollutants. Further work is needed to improve the efficiency and cost-effectiveness of these devices for industrial use.

SOLAR THERMAL ENERGY

Solar thermal energy (STE) is a form of energy and a technology for harnessing solar energy to generate thermal energy or electrical energy for use in industry, and in the residential and commercial sectors.

Solar thermal collectors are classified by the United States Energy Information Administration as low-, medium-, or high-temperature collectors. Low-temperature collectors are generally unglazed and used to heat swimming pools or to heat ventilation air. Medium-temperature collectors are also usually flat plates but are used for heating water or air for residential and commercial use. High-temperature collectors concentrate sunlight using mirrors or lenses and are generally used for fulfilling heat requirements up to 300 deg C/20 bar pressure in industries, and for electric power production. Two categories include Concentrated Solar Thermal (CST) for fulfilling heat requirements in industries, and Concentrated Solar Power (CSP) when the heat collected is used for power generation. CST and CSP are not replaceable in terms of application. The largest facilities are located in the American Mojave Desert of California and Nevada. These plants employ a variety of different technologies. The largest examples include, Ivanpah Solar Power Facility (377 MW), Solar Energy Generating Systems installation (354 MW), and Crescent Dunes (110 MW). Spain is the other major developer of solar thermal power plant. The largest examples include, Solnova Solar Power Station (150 MW), the Andasol solar power station (150 MW), and Extresol Solar Power Station (100 MW).

Low-Temperature Solar Heating and Cooling Systems

MIT's Solar House built in 1939 used seasonal thermal energy
storage (STES) for year-round heating.

Systems for utilizing low-temperature solar thermal energy include means for heat collection; usually heat storage, either short-term or interseasonal; and distribution within a structure or a district heating network. In some cases more than one of these functions is inherent to a single feature of the system (e.g. some kinds of solar collectors also store heat). Some systems are passive, others are active (requiring other external energy to function).

Heating is the most obvious application, but solar cooling can be achieved for a building or district cooling network by using a heat-driven absorption or adsorption chiller (heat pump). There is a productive coincidence that the greater the driving heat from insolation, the greater the cooling output. In 1878, Auguste Mouchout pioneered solar cooling by making ice using a solar steam engine attached to a refrigeration device.

In the United States, heating, ventilation, and air conditioning (HVAC) systems account for over 25% (4.75 EJ) of the energy used in commercial buildings (50% in northern cities) and nearly half (10.1 EJ) of the energy used in residential buildings. Solar heating, cooling, and ventilation technologies can be used to offset a portion of this energy.

The most popular solar heating technology for heating buildings is the building integrated transpired solar air collection system which connects to the building's HVAC equipment. According to Solar Energy Industries Association over 500,000 m² (5,000,000 square feet) of these panels are in operation in North America as of 2015.

In Europe, since the mid-1990s about 125 large solar-thermal district heating plants have been constructed, each with over 500 m² (5400 ft²) of solar collectors. The largest are about 10,000 m², with capacities of 7 MW-thermal and solar heat costs around 4 Eurocents/kWh without subsidies. 40 of them have nominal capacities of 1 MW-thermal or more. The Solar District Heating program (SDH) has participation from 14 European Nations and the European Commission, and is working toward technical and market development, and holds annual conferences.

Low-Temperature Collectors

Glazed solar collectors are designed primarily for space heating. They recirculate building air through a solar air panel where the air is heated and then directed back into the building. These solar space heating systems require at least two penetrations into the building and only perform when the air in the solar collector is warmer than the building room temperature. Most glazed collectors are used in the residential sector.

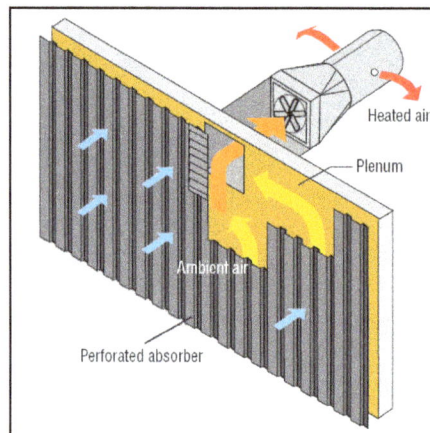

Unglazed, "transpired" air collector.

Unglazed solar collectors are primarily used to pre-heat make-up ventilation air in commercial, industrial and institutional buildings with a high ventilation load. They turn building walls or sections of walls into low cost, high performance, unglazed solar collectors. Also called, "transpired solar panels" or "solar wall", they employ a painted perforated metal solar heat absorber that also serves as the exterior wall surface of the building. Heat transfer to the air takes place on the surface of the absorber, through the metal absorber and behind the absorber. The boundary layer of solar heated air is drawn into a nearby perforation before the heat can escape by convection to the outside air. The heated air is then drawn from behind the absorber plate into the building's ventilation system.

Building integrated unglazed transpired solar air collector with grey walls
and white canopy/collection ducts.

A Trombe wall is a passive solar heating and ventilation system consisting of an air channel sandwiched between a window and a sun-facing thermal mass. During the ventilation cycle, sunlight stores heat in the thermal mass and warms the air channel causing circulation through vents at the top and bottom of the wall. During the heating cycle the Trombe wall radiates stored heat.

Solar roof ponds are unique solar heating and cooling systems developed by Harold Hay in the 1960s. A basic system consists of a roof-mounted water bladder with a movable insulating cover. This system can control heat exchange between interior and exterior environments by covering and uncovering the bladder between night and day. When heating is a concern the bladder is uncovered during the day allowing sunlight to warm the water bladder and store heat for evening use. When cooling is a concern the covered bladder draws heat from the building's interior during the day and is uncovered at night to radiate heat to the cooler atmosphere. The Skytherm house in Atascadero, California uses a prototype roof pond for heating and cooling.

Solar space heating with solar air heat collectors is more popular in the USA and Canada than heating with solar liquid collectors since most buildings already have a ventilation system for heating and cooling. The two main types of solar air panels are glazed and unglazed.

Of the 21,000,000 square feet (2,000,000 m²) of solar thermal collectors produced in the United States in 2007, 16,000,000 square feet (1,500,000 m²) were of the low-temperature variety. Low-temperature collectors are generally installed to heat swimming pools, although they can also be used for space heating. Collectors can use air or water as the medium to transfer the heat to their destination.

Heat Storage in Low-Temperature Solar Thermal Systems

Interseasonal storage. Solar heat (or heat from other sources) can be effectively stored between opposing seasons in aquifers, underground geological strata, large specially constructed pits, and large tanks that are insulated and covered with earth.

Short-term storage. Thermal mass materials store solar energy during the day and release this energy during cooler periods. Common thermal mass materials include stone, concrete, and water. The proportion and placement of thermal mass should consider several factors such as climate, daylighting, and shading conditions. When properly incorporated, thermal mass can passively maintain comfortable temperatures while reducing energy consumption.

Solar-Driven Cooling

Worldwide, by 2011 there were about 750 cooling systems with solar-driven heat pumps, and annual market growth was 40 to 70% over the prior seven years. It is a niche market because the economics are challenging, with the annual number of cooling hours a limiting factor. Respectively, the annual cooling hours are roughly 1000 in the Mediterranean, 2500 in Southeast Asia, and only 50 to 200 in Central Europe. However, system construction costs dropped about 50% between 2007 and 2011. The International Energy Agency (IEA) Solar Heating and Cooling program (IEA-SHC) task groups working on further development of the technologies involved.

Solar Heat-Driven Ventilation

A solar chimney (or thermal chimney) is a passive solar ventilation system composed of a hollow thermal mass connecting the interior and exterior of a building. As the chimney warms, the air inside is heated causing an updraft that pulls air through the building. These systems have been in use since Roman times and remain common in the Middle East.

Process Heat

Solar Evaporation Ponds in the Atacama Desert.

Solar process heating systems are designed to provide large quantities of hot water or space heating for nonresidential buildings.

Evaporation ponds are shallow ponds that concentrate dissolved solids through evaporation. The use of evaporation ponds to obtain salt from sea water is one of the oldest applications of solar energy. Modern uses include concentrating brine solutions used in leach mining and removing dissolved solids from waste streams. Altogether, evaporation ponds represent one of the largest commercial applications of solar energy in use today.

Unglazed transpired collectors are perforated sun-facing walls used for preheating ventilation air. Transpired collectors can also be roof mounted for year-round use and can raise the incoming air temperature up to 22 °C and deliver outlet temperatures of 45-60 °C. The short payback period of transpired collectors (3 to 12 years) make them a more cost-effective alternative to glazed collection systems. As of 2015, over 4000 systems with a combined collector area of 500,000 m² had been installed worldwide. Representatives include an 860 m² collector in Costa Rica used for drying coffee beans and a 1300 m² collector in Coimbatore, India used for drying marigolds.

A food processing facility in Modesto, California uses parabolic troughs to produce steam used in the manufacturing process. The 5,000 m² collector area is expected to provide 15 TJ per year.

Medium-Temperature Collectors

These collectors could be used to produce approximately 50% and more of the hot water needed for residential and commercial use in the United States. In the United States, a typical system costs $4000–$6000 retail ($1400 to $2200 wholesale for the materials) and 30% of the system qualifies for a federal tax credit + additional state credit exists in about half of the states. Labor for a simple open loop system in southern climates can take 3–5 hours for the installation and 4–6 hours in Northern areas. Northern system require more collector area and more complex plumbing to protect the collector from freezing. With this incentive, the payback time for a typical household is four to nine years, depending on the state. Similar subsidies exist in parts of Europe. A crew of one solar plumber and two assistants with minimal training can install a system per day. Thermosiphon installation have negligible maintenance costs (costs rise if antifreeze and mains power are used for circulation) and in the US reduces a households' operating costs by $6 per person per month. Solar water heating can reduce CO_2 emissions of a family of four by 1 ton/year (if replacing natural gas) or 3 ton/year (if replacing electricity). Medium-temperature installations can use any of several designs: common designs are pressurized glycol, drain back, batch systems and newer low pressure freeze tolerant systems using polymer pipes containing water with photovoltaic pumping. European and International standards are being reviewed to accommodate innovations in design and operation of medium temperature collectors. Operational innovations include "permanently wetted collector" operation. This innovation reduces or even eliminates the occurrence of no-flow high temperature stresses called stagnation which would otherwise reduce the life expectancy of collectors.

Solar Drying

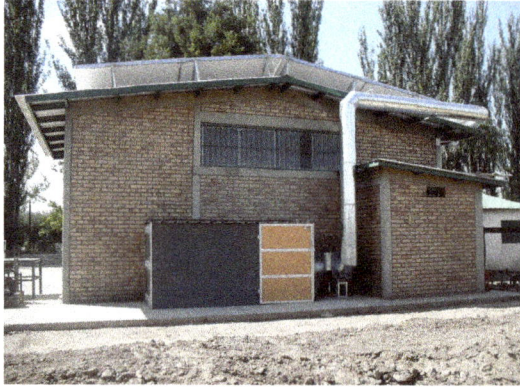

Industrial indirect solar fruit and vegetable dryer.

Solar thermal energy can be useful for drying wood for construction and wood fuels such as wood chips for combustion. Solar is also used for food products such as fruits, grains, and fish. Crop drying by solar means is environmentally friendly as well as cost effective while improving the quality. The less money it takes to make a product, the less it can be sold for, pleasing both the buyers and the sellers. Technologies in solar drying include ultra low cost pumped transpired plate air collectors based on black fabrics. Solar thermal energy is helpful in the process of drying products such as wood chips and other forms of biomass by raising the temperature while allowing air to pass through and get rid of the moisture.

Cooking

The Solar Bowl above the Solar Kitchen in Auroville, India concentrates sunlight on a movable receiver to produce steam for cooking.

Solar cookers use sunlight for cooking, drying and pasteurization. Solar cooking offsets fuel costs, reduces demand for fuel or firewood, and improves air quality by reducing or removing a source of smoke.

The simplest type of solar cooker is the box cooker first built by Horace de Saussure in 1767. A basic box cooker consists of an insulated container with a transparent lid. These

These cookers can be used effectively with partially overcast skies and will typically reach temperatures of 50–100 °C.

Concentrating solar cookers use reflectors to concentrate solar energy onto a cooking container. The most common reflector geometries are flat plate, disc and parabolic trough type. These designs cook faster and at higher temperatures (up to 350 °C) but require direct light to function properly.

The Solar Kitchen in Auroville, India uses a unique concentrating technology known as the solar bowl. Contrary to conventional tracking reflector/fixed receiver systems, the solar bowl uses a fixed spherical reflector with a receiver which tracks the focus of light as the Sun moves across the sky. The solar bowl's receiver reaches temperature of 150 °C that is used to produce steam that helps cook 2,000 daily meals.

Many other solar kitchens in India use another unique concentrating technology known as the Scheffler reflector. This technology was first developed by Wolfgang Scheffler in 1986. A Scheffler reflector is a parabolic dish that uses single axis tracking to follow the Sun's daily course. These reflectors have a flexible reflective surface that is able to change its curvature to adjust to seasonal variations in the incident angle of sunlight. Scheffler reflectors have the advantage of having a fixed focal point which improves the ease of cooking and are able to reach temperatures of 450-650 °C. Built in 1999 by the Brahma Kumaris, the world's largest Scheffler reflector system in Abu Road, Rajasthan India is capable of cooking up to 35,000 meals a day. By early 2008, over 2000 large cookers of the Scheffler design had been built worldwide.

Distillation

Solar stills can be used to make drinking water in areas where clean water is not common. Solar distillation is necessary in these situations to provide people with purified water. Solar energy heats up the water in the still. The water then evaporates and condenses on the bottom of the covering glass.

High-Temperature Collectors

Part of the 354 MW SEGS solar complex in northern San Bernardino County, California.

The solar furnace at Odeillo in the French Pyrenees-Orientales
can reach temperatures up to 3,500 °C .

Where temperatures below about 95 °C are sufficient, as for space heating, flat-plate collectors of the nonconcentrating type are generally used. Because of the relatively high heat losses through the glazing, flat plate collectors will not reach temperatures much above 200 °C even when the heat transfer fluid is stagnant. Such temperatures are too low for efficient conversion to electricity.

The efficiency of heat engines increases with the temperature of the heat source. To achieve this in solar thermal energy plants, solar radiation is concentrated by mirrors or lenses to obtain higher temperatures – a technique called Concentrated Solar Power (CSP). The practical effect of high efficiencies is to reduce the plant's collector size and total land use per unit power generated, reducing the environmental impacts of a power plant as well as its expense.

As the temperature increases, different forms of conversion become practical. Up to 600 °C, steam turbines, standard technology, have an efficiency up to 41%. Above 600 °C, gas turbines can be more efficient. Higher temperatures are problematic because different materials and techniques are needed. One proposal for very high temperatures is to use liquid fluoride salts operating between 700 °C to 800 °C, using multistage turbine systems to achieve 50% or more thermal efficiencies. The higher operating temperatures permit the plant to use higher-temperature dry heat exchangers for its thermal exhaust, reducing the plant's water use – critical in the deserts where large solar plants are practical. High temperatures also make heat storage more efficient, because more watt-hours are stored per unit of fluid.

Commercial concentrating solar thermal power (CSP) plants were first developed in the 1980s. The world's largest solar thermal power plants are now the 370 MW Ivanpah Solar Power Facility, commissioned in 2014, and the 354 MW SEGS CSP installation, both located in the Mojave Desert of California, where several other solar projects have been realized as well. With the exception of the Shams solar power station, built in 2013 near Abu Dhabi, the United Arab Emirates, all other 100 MW or larger CSP plants are either located in the United States or in Spain.

The principal advantage of CSP is the ability to efficiently add thermal storage, allowing the dispatching of electricity over up to a 24-hour period. Since peak electricity demand typically occurs between about 4 and 8 pm, many CSP power plants use 3 to 5 hours of thermal storage. With current technology, storage of heat is much cheaper and more efficient than storage of electricity. In this way, the CSP plant can produce electricity day and night. If the CSP site has predictable solar radiation, then the CSP plant becomes a reliable power plant. Reliability can further be improved by installing a back-up combustion system. The back-up system can use most of the CSP plant, which decreases the cost of the back-up system.

CSP facilities utilize high electrical conductivity materials, such as copper, in field power cables, grounding networks, and motors for tracking and pumping fluids, as well as in the main generator and high voltage transformers.

With reliability, unused desert, no pollution, and no fuel costs, the obstacles for large deployment for CSP are cost, aesthetics, land use and similar factors for the necessary connecting high tension lines. Although only a small percentage of the desert is necessary to meet global electricity demand, still a large area must be covered with mirrors or lenses to obtain a significant amount of energy. An important way to decrease cost is the use of a simple design.

When considering land use impacts associated with the exploration and extraction through to transportation and conversion of fossil fuels, which are used for most of our electrical power, utility-scale solar power compares as one of the most land-efficient energy resources available:

The federal government has dedicated nearly 2,000 times more acreage to oil and gas leases than to solar development. In 2010 the Bureau of Land Management approved nine large-scale solar projects, with a total generating capacity of 3,682 megawatts, representing approximately 40,000 acres. In contrast, in 2010, the Bureau of Land Management processed more than 5,200 applications gas and oil leases, and issued 1,308 leases, for a total of 3.2 million acres. Currently, 38.2 million acres of onshore public lands and an additional 36.9 million acres of offshore exploration in the Gulf of Mexico are under lease for oil and gas development, exploration and production.

System Designs

During the day the sun has different positions. For low concentration systems (and low temperatures) tracking can be avoided (or limited to a few positions per year) if nonimaging optics are used. For higher concentrations, however, if the mirrors or lenses do not move, then the focus of the mirrors or lenses changes (but also in these cases nonimaging optics provides the widest acceptance angles for a given concentration). Therefore, it seems unavoidable that there needs to be a tracking system that follows the position of the sun (for solar photovoltaic a solar tracker is only optional).

The tracking system increases the cost and complexity. With this in mind, different designs can be distinguished in how they concentrate the light and track the position of the sun.

Parabolic Trough Designs

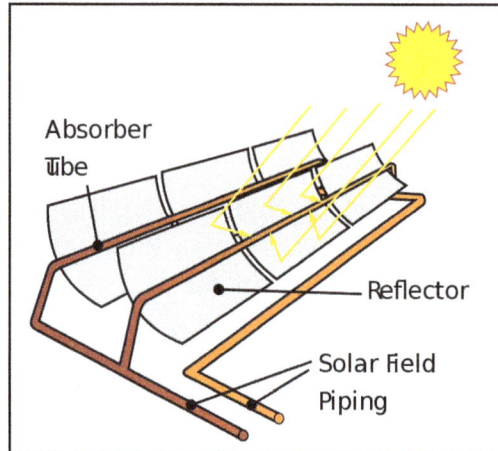

Sketch of a parabolic trough design. A change of position of the sun parallel to the receiver does not require adjustment of the mirrors.

Parabolic trough power plants use a curved, mirrored trough which reflects the direct solar radiation onto a glass tube containing a fluid (also called a receiver, absorber or collector) running the length of the trough, positioned at the focal point of the reflectors. The trough is parabolic along one axis and linear in the orthogonal axis. For change of the daily position of the sun perpendicular to the receiver, the trough tilts east to west so that the direct radiation remains focused on the receiver. However, seasonal changes in the angle of sunlight parallel to the trough does not require adjustment of the mirrors, since the light is simply concentrated elsewhere on the receiver. Thus the trough design does not require tracking on a second axis. The receiver may be enclosed in a glass vacuum chamber. The vacuum significantly reduces convective heat loss.

A fluid (also called heat transfer fluid) passes through the receiver and becomes very hot. Common fluids are synthetic oil, molten salt and pressurized steam. The fluid containing the heat is transported to a heat engine where about a third of the heat is converted to electricity.

Full-scale parabolic trough systems consist of many such troughs laid out in parallel over a large area of land. Since 1985 a solar thermal system using this principle has been in full operation in California in the United States. It is called the Solar Energy Generating Systems (SEGS) system. Other CSP designs lack this kind of long experience and therefore it can currently be said that the parabolic trough design is the most thoroughly proven CSP technology.

The SEGS is a collection of nine plants with a total capacity of 354 MW and has been the world's largest solar power plant, both thermal and non-thermal, for many years.

A newer plant is Nevada Solar One plant with a capacity of 64 MW. The 150 MW Andasol solar power stations are in Spain with each site having a capacity of 50 MW. Note however, that those plants have heat storage which requires a larger field of solar collectors relative to the size of the steam turbine-generator to store heat and send heat to the steam turbine at the same time. Heat storage enables better utilization of the steam turbine. With day and some nighttime operation of the steam-turbine Andasol 1 at 50 MW peak capacity produces more energy than Nevada Solar One at 64 MW peak capacity, due to the former plant's thermal energy storage system and larger solar field. The 280MW Solana Generating Station came online in Arizona in 2013 with 6 hours of power storage. Hassi R'Mel integrated solar combined cycle power station in Algeria and Martin Next Generation Solar Energy Center both use parabolic troughs in a combined cycle with natural gas.

Enclosed Trough

Inside an enclosed trough system.

The enclosed trough architecture encapsulates the solar thermal system within a greenhouse-like glasshouse. The glasshouse creates a protected environment to withstand the elements that can negatively impact reliability and efficiency of the solar thermal system.

Lightweight curved solar-reflecting mirrors are suspended within the glasshouse structure. A single-axis tracking system positions the mirrors to track the sun and focus its light onto a network of stationary steel pipes, also suspended from the glasshouse structure. Steam is generated directly, using oil field-quality water, as water flows from the inlet throughout the length of the pipes, without heat exchangers or intermediate working fluids.

The steam produced is then fed directly to the field's existing steam distribution network, where the steam is continuously injected deep into the oil reservoir. Sheltering the mirrors from the wind allows them to achieve higher temperature rates and prevents dust from building up as a result from exposure to humidity. GlassPoint Solar, the company that created the Enclosed Trough design, states its technology can produce heat for EOR for about $5 per million British thermal units in sunny regions, compared to between $10 and $12 for other conventional solar thermal technologies.

GlassPoint's enclosed trough system has been utilized at the Miraah facility in Oman, and a new project has recently been announced for the company to bring its enclosed trough technology to the South Belridge Oil Field, near Bakersfield, California.

Power Tower Designs

Ivanpah Solar Electric Generating System with all three towers under load,
The Clark Mountain Range can be seen in the distance.

Power towers (also known as 'central tower' power plants or 'heliostat' power plants) capture and focus the sun's thermal energy with thousands of tracking mirrors (called heliostats) in roughly a two square mile field. A tower resides in the center of the heliostat field. The heliostats focus concentrated sunlight on a receiver which sits on top of the tower. Within the receiver the concentrated sunlight heats molten salt to over 1,000 °F (538 °C). The heated molten salt then flows into a thermal storage tank where it is stored, maintaining 98% thermal efficiency, and eventually pumped to a steam generator. The steam drives a standard turbine to generate electricity. This process, also known as the "Rankine cycle" is similar to a standard coal-fired power plant, except it is fueled by clean and free solar energy.

The advantage of this design above the parabolic trough design is the higher temperature. Thermal energy at higher temperatures can be converted to electricity more efficiently and can be more cheaply stored for later use. Furthermore, there is less need to flatten the ground area. In principle a power tower can be built on the side of a hill. Mirrors can be flat and plumbing is concentrated in the tower. The disadvantage is that each mirror must have its own dual-axis control, while in the parabolic trough design single axis tracking can be shared for a large array of mirrors.

A cost/performance comparison between power tower and parabolic trough concentrators was made by the NREL which estimated that by 2020 electricity could be produced from power towers for 5.47 ¢/kWh and for 6.21 ¢/kWh from parabolic troughs.

The capacity factor for power towers was estimated to be 72.9% and 56.2% for parabolic troughs. There is some hope that the development of cheap, durable, mass producible heliostat power plant components could bring this cost down.

The first commercial tower power plant was PS10 in Spain with a capacity of 11 MW, completed in 2007. Since then a number of plants have been proposed, several have been built in a number of countries (Spain, Germany, U.S. Turkey, China, India) but several proposed plants were cancelled as photovoltaic solar prices plummeted. A solar power tower went online in South Africa in 2016. Ivanpah Solar Power Facility in California generates 392 MW of electricity from three towers, making it the largest solar power tower plant when it came online in late 2013.

Dish Designs

A parabolic solar dish concentrating the sun's rays on the heating element of a Stirling engine. The entire unit acts as a solar tracker.

CSP-Stirling is known to have the highest efficiency of all solar technologies (around 30%, compared to solar photovoltaic's approximately 15%), and is predicted to be able to produce the cheapest energy among all renewable energy sources in high-scale production and hot areas, semi-deserts, etc. A dish Stirling system uses a large, reflective, parabolic dish (similar in shape to a satellite television dish). It focuses all the sunlight that strikes the dish up onto a single point above the dish, where a receiver captures the heat and transforms it into a useful form. Typically the dish is coupled with a Stirling engine in a Dish-Stirling System, but also sometimes a steam engine is used. These create rotational kinetic energy that can be converted to electricity using an electric generator.

In 2005 Southern California Edison announced an agreement to purchase solar powered Stirling engines from Stirling Energy Systems over a twenty-year period and in quantities (20,000 units) sufficient to generate 500 megawatts of electricity. In January 2010, Stirling Energy Systems and Tessera Solar commissioned the first demonstration 1.5-megawatt power plant ("Maricopa Solar") using Stirling technology in Peoria, Arizona. At the beginning of 2011 Stirling Energy's development arm, Tessera Solar, sold off its two large projects, the 709 MW Imperial project and the 850 MW Calico project to AES Solar and K.Road, respectively. In 2012 the Maricopa plant was bought and dismantled by United Sun Systems. United Sun Systems released a new generation system, based on a V-shaped Stirling engine and a peak production of 33 kW. The new CSP-Stirling technology brings down LCOE to USD 0.02 in utility scale.

According to its developer, Rispasso Energy, a Swedish firm, in 2015 its Dish Sterling system being tested in the Kalahari Desert in South Africa showed 34% efficiency.

Fresnel Technologies

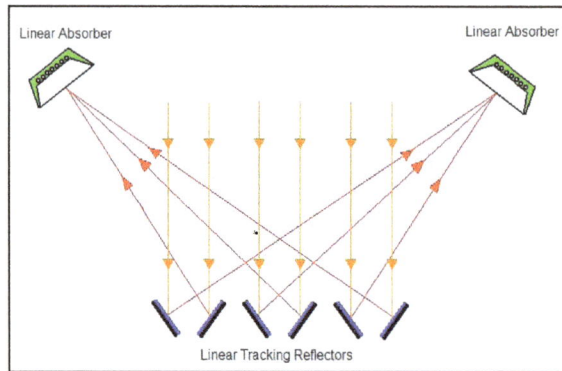

Fresnel reflector.

A linear Fresnel reflector power plant uses a series of long, narrow, shallow-curvature (or even flat) mirrors to focus light onto one or more linear receivers positioned above the mirrors. On top of the receiver a small parabolic mirror can be attached for further focusing the light. These systems aim to offer lower overall costs by sharing a receiver between several mirrors (as compared with trough and dish concepts), while still using the simple line-focus geometry with one axis for tracking. This is similar to the trough design (and different from central towers and dishes with dual-axis). The receiver is stationary and so fluid couplings are not required (as in troughs and dishes). The mirrors also do not need to support the receiver, so they are structurally simpler. When suitable aiming strategies are used (mirrors aimed at different receivers at different times of day), this can allow a denser packing of mirrors on available land area.

Rival single axis tracking technologies include the relatively new linear Fresnel reflector (LFR) and compact-LFR (CLFR) technologies. The LFR differs from that of the parabolic trough in that the absorber is fixed in space above the mirror field. Also, the reflector is composed of many low row segments, which focus collectively on an elevated long tower receiver running parallel to the reflector rotational axis.

Prototypes of Fresnel lens concentrators have been produced for the collection of thermal energy by International Automated Systems. No full-scale thermal systems using Fresnel lenses are known to be in operation, although products incorporating Fresnel lenses in conjunction with photovoltaic cells are already available.

MicroCSP

MicroCSP is used for community-sized power plants (1 MW to 50 MW), for industrial, agricultural and manufacturing 'process heat' applications, and when large amounts of hot water are needed, such as resort swimming pools, water parks, large laundry facilities, sterilization, distillation and other such uses.

Enclosed Parabolic Trough

The enclosed parabolic trough solar thermal system encapsulates the components within an off-the-shelf greenhouse type of glasshouse. The glasshouse protects the components from the elements that can negatively impact system reliability and efficiency. This protection importantly includes nightly glass-roof washing with optimized water-efficient off-the-shelf automated washing systems. Lightweight curved solar-reflecting mirrors are suspended from the ceiling of the glasshouse by wires. A single-axis tracking system positions the mirrors to retrieve the optimal amount of sunlight. The mirrors concentrate the sunlight and focus it on a network of stationary steel pipes, also suspended from the glasshouse structure. Water is pumped through the pipes and boiled to generate steam when intense sun radiation is applied. The steam is available for process heat. Sheltering the mirrors from the wind allows them to achieve higher temperature rates and prevents dust from building up on the mirrors as a result from exposure to humidity.

Heat Collection and Exchange

More energy is contained in higher frequency light based upon the formula of $E = h\nu$, where h is the Planck constant and ν is frequency. Metal collectors down convert higher frequency light by producing a series of Compton shifts into an abundance of lower frequency light. Glass or ceramic coatings with high transmission in the visible and UV and effective absorption in the IR (heat blocking) trap metal absorbed low frequency light from radiation loss. Convection insulation prevents mechanical losses transferred through gas. Once collected as heat, thermos containment efficiency improves significantly with increased size. Unlike Photovoltaic technologies that often degrade under concentrated light, Solar Thermal depends upon light concentration that requires a clear sky to reach suitable temperatures.

Heat in a solar thermal system is guided by five basic principles: heat gain; heat transfer; heat storage; heat transport; and heat insulation. Here, heat is the measure of the amount of thermal energy an object contains and is determined by the temperature, mass and specific heat of the object. Solar thermal power plants use heat exchangers that are designed for constant working conditions, to provide heat exchange. Copper heat exchangers are important in solar thermal heating and cooling systems because of copper's high thermal conductivity, resistance to atmospheric and water corrosion, sealing and joining by soldering, and mechanical strength. Copper is used both in receivers and in primary circuits (pipes and heat exchangers for water tanks) of solar thermal water systems.

Heat gain is the heat accumulated from the sun in the system. Solar thermal heat is trapped using the greenhouse effect; the greenhouse effect in this case is the ability of a reflective surface to transmit short wave radiation and reflect long wave radiation. Heat and infrared radiation (IR) are produced when short wave radiation light hits the absorber plate, which is then trapped inside the collector. Fluid, usually water, in the absorber tubes collect the trapped heat and transfer it to a heat storage vault.

Heat is transferred either by conduction or convection. When water is heated, kinetic energy is transferred by conduction to water molecules throughout the medium. These molecules spread their thermal energy by conduction and occupy more space than the cold slow moving molecules above them. The distribution of energy from the rising hot water to the sinking cold water contributes to the convection process. Heat is transferred from the absorber plates of the collector in the fluid by conduction. The collector fluid is circulated through the carrier pipes to the heat transfer vault. Inside the vault, heat is transferred throughout the medium through convection.

Heat storage enables solar thermal plants to produce electricity during hours without sunlight. Heat is transferred to a thermal storage medium in an insulated reservoir during hours with sunlight, and is withdrawn for power generation during hours lacking sunlight. Thermal storage mediums will be discussed in a heat storage section. Rate of heat transfer is related to the conductive and convection medium as well as the temperature differences. Bodies with large temperature differences transfer heat faster than bodies with lower temperature differences.

Heat transport refers to the activity in which heat from a solar collector is transported to the heat storage vault. Heat insulation is vital in both heat transport tubing as well as the storage vault. It prevents heat loss, which in turn relates to energy loss, or decrease in the efficiency of the system.

Heat Storage for Space Heating

A collection of mature technologies called seasonal thermal energy storage (STES) is capable of storing heat for months at a time, so solar heat collected primarily in Summer can be used for all-year heating. Solar-supplied STES technology has been advanced primarily in Denmark, Germany, and Canada, and applications include individual buildings and district heating networks. Drake Landing Solar Community in Alberta, Canada has a small district system and in 2012 achieved a world record of providing 97% of the community's all-year space heating needs from the sun. STES thermal storage mediums include deep aquifers; native rock surrounding clusters of small-diameter, heat exchanger equipped boreholes; large, shallow, lined pits that are filled with gravel and top-insulated; and large, insulated and buried surface water tanks.

Heat Storage to Stabilize Solar-Electric Power Generation

Heat storage allows a solar thermal plant to produce electricity at night and on overcast days. This allows the use of solar power for baseload generation as well as peak power generation, with the potential of displacing both coal- and natural gas-fired power plants. Additionally, the utilization of the generator is higher which reduces cost. Even short term storage can help by smoothing out the "duck curve" of rapid change in generation requirements at sunset when a grid includes large amounts of solar capacity.

Heat is transferred to a thermal storage medium in an insulated reservoir during the day, and withdrawn for power generation at night. Thermal storage media include pressurized steam, concrete, a variety of phase change materials, and molten salts such as calcium, sodium and potassium nitrate.

Steam Accumulator

The PS10 solar power tower stores heat in tanks as pressurized steam at 50 bar and 285 °C. The steam condenses and flashes back to steam, when pressure is lowered. Storage is for one hour. It is suggested that longer storage is possible, but that has not been proven in an existing power plant.

Molten Salt Storage

The 150 MW Andasol solar power station is a commercial parabolic trough solar thermal power plant, located in Spain. The Andasol plant uses tanks of molten salt to store solar energy so that it can continue generating electricity even when the sun isn't shining.

A variety of fluids have been tested to transport the sun's heat, including water, air, oil, and sodium, but Rockwell International selected molten salt as best. Molten salt is used in solar power tower systems because it is liquid at atmospheric pressure, provides a low-cost medium to store thermal energy, its operating temperatures are compatible with today's steam turbines, and it is non-flammable and nontoxic. Molten salt is used in the chemical and metals industries to transport heat, so industry has experience with it.

The first commercial molten salt mixture was a common form of saltpeter, 60% sodium nitrate and 40% potassium nitrate. Saltpeter melts at 220 °C (430 °F) and is kept liquid at 290 °C (550 °F) in an insulated storage tank. Calcium nitrate can reduce the melting point to 131 °C, permitting more energy to be extracted before the salt freezes. There are now several technical calcium nitrate grades stable at more than 500 °C.

This solar power system can generate power in cloudy weather or at night using the heat in the tank of hot salt. The tanks are insulated, able to store heat for a week. Tanks that power a 100-megawatt turbine for four hours would be about 9 m (30 ft) tall and 24 m (80 ft) in diameter.

The Andasol power plant in Spain is the first commercial solar thermal power plant using molten salt for heat storage and nighttime generation. It came on line March 2009. On July 4, 2011, a company in Spain celebrated an historic moment for the solar industry: Torresol's 19.9 MW concentrating solar power plant became the first ever to generate uninterrupted electricity for 24 hours straight, using a molten salt heat storage.

In 2016 SolarReserve proposed a 2 GW, $5 billion concentrated solar plant with storage in Nevada.

In January 2019 Shouhang Energy Saving Dunhuang 100MW molten salt tower solar energy photothermal power station project was connected to grid and started operating. Its configuration includes an 11-hour molten salt heat storage system and can generate power consecutively for 24 hours.

Phase-Change Materials for Storage

Phase Change Material (PCMs) offer an alternative solution in energy storage. Using a similar heat transfer infrastructure, PCMs have the potential of providing a more efficient means of storage. PCMs can be either organic or inorganic materials. Advantages of organic PCMs include no corrosives, low or no undercooling, and chemical and thermal stability. Disadvantages include low phase-change enthalpy, low thermal conductivity, and flammability. Inorganics are advantageous with greater phase-change enthalpy, but exhibit disadvantages with undercooling, corrosion, phase separation, and lack of thermal stability. The greater phase-change enthalpy in inorganic PCMs make hydrate salts a strong candidate in the solar energy storage field.

Use of Water

A design which requires water for condensation or cooling may conflict with location of solar thermal plants in desert areas with good solar radiation but limited water resources. The conflict is illustrated by plans of Solar Millennium, a German company, to build a plant in the Amargosa Valley of Nevada which would require 20% of the water available in the area. Some other projected plants by the same and other companies in the Mojave Desert of California may also be affected by difficulty in obtaining adequate and appropriate water rights. California water law currently prohibits use of potable water for cooling.

Other designs require less water. The Ivanpah Solar Power Facility in south-eastern California conserves scarce desert water by using air-cooling to convert the steam back into water. Compared to conventional wet-cooling, this results in a 90% reduction in water usage at the cost of some loss of efficiency. The water is then returned to the boiler in a closed process which is environmentally friendly.

Conversion Rates from Solar Energy to Electrical Energy

Of all of these technologies the solar dish/Stirling engine has the highest energy efficiency. A single solar dish-Stirling engine installed at Sandia National Laboratories National Solar Thermal Test Facility (NSTTF) produces as much as 25 kW of electricity, with a conversion efficiency of 31.25%.

Solar parabolic trough plants have been built with efficiencies of about 20%. Fresnel reflectors have an efficiency that is slightly lower (but this is compensated by the denser packing).

The gross conversion efficiencies (taking into account that the solar dishes or troughs occupy only a fraction of the total area of the power plant) are determined by net generating capacity over the solar energy that falls on the total area of the solar plant. The 500-megawatt (MW) SCE/SES plant would extract about 2.75% of the radiation that falls on its 4,500 acres (18.2 km²). For the 50 MW AndaSol Power Plant that is being built in Spain (total area of 1,300×1,500 m = 1.95 km²) gross conversion efficiency comes out at 2.6%.

Furthermore, efficiency does not directly relate to cost: on calculating total cost, both efficiency and the cost of construction and maintenance should be taken into account.

SOLAR POWER

Solar power is the conversion of energy from sunlight into electricity, either directly using photovoltaics (PV), indirectly using concentrated solar power, or a combination. Concentrated solar power systems use lenses or mirrors and tracking systems to focus a large area of sunlight into a small beam. Photovoltaic cells convert light into an electric current using the photovoltaic effect.

Photovoltaics were initially solely used as a source of electricity for small and medium-sized applications, from the calculator powered by a single solar cell to remote homes powered by an off-grid rooftop PV system. Commercial concentrated solar power plants were first developed in the 1980s. The 392 MW Ivanpah installation is the largest concentrating solar power plant in the world, located in the Mojave Desert of California.

As the cost of solar electricity has fallen, the number of grid-connected solar PV systems has grown into the millions and utility-scale photovoltaic power stations with hundreds of megawatts are being built. Solar PV is rapidly becoming an inexpensive, low-carbon technology to harness renewable energy from the Sun. The current largest photovoltaic power station in the world is the 850 MW Longyangxia Dam Solar Park, in Qinghai, China.

The International Energy Agency projected in 2014 that under its "high renewables" scenario, by 2050, solar photovoltaics and concentrated solar power would contribute about 16 and 11 percent, respectively, of the worldwide electricity consumption, and solar would be the world's largest source of electricity. Most solar installations would be in China and India. In 2017, solar power provided 1.7% of total worldwide electricity production, growing at 35% per annum. As of 2018, the unsubsidised levelised cost of electricity for utility scale solar power is around $43/MWh.

Mainstream Technologies

Many industrialized nations have installed significant solar power capacity into their grids to supplement or provide an alternative to conventional energy sources while an increasing number of less developed nations have turned to solar to reduce dependence on expensive imported fuels. Long distance transmission allows remote renewable energy resources to displace fossil fuel consumption. Solar power plants use one of two technologies:

- Photovoltaic (PV) systems use solar panels, either on rooftops or in ground-mounted solar farms, converting sunlight directly into electric power.

- Concentrated solar power (CSP, also known as "concentrated solar thermal") plants use solar thermal energy to make steam, that is thereafter converted into electricity by a turbine.

Photovoltaics

A solar cell, or photovoltaic cell (PV), is a device that converts light into electric current using the photovoltaic effect. The first solar cell was constructed by Charles Fritts in the 1880s. The German industrialist Ernst Werner von Siemens was among those who recognized the importance of this discovery. In 1931, the German engineer Bruno Lange developed a photo cell using silver selenide in place of copper oxide, although the prototype selenium cells converted less than 1% of incident light into electricity. Following the work of Russell Ohl in the 1940s, researchers Gerald Pearson, Calvin Fuller and

Daryl Chapin created the silicon solar cell in 1954. These early solar cells cost US$286/watt and reached efficiencies of 4.5–6%.

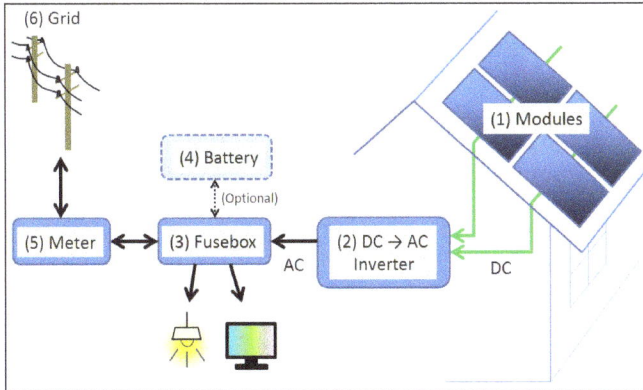

Schematics of a grid-connected residential PV power system.

The array of a photovoltaic power system, or PV system, produces direct current (DC) power which fluctuates with the sunlight's intensity. For practical use this usually requires conversion to certain desired voltages or alternating current (AC), through the use of inverters. Multiple solar cells are connected inside modules. Modules are wired together to form arrays, then tied to an inverter, which produces power at the desired voltage, and for AC, the desired frequency/phase.

Many residential PV systems are connected to the grid wherever available, especially in developed countries with large markets. In these grid-connected PV systems, use of energy storage is optional. In certain applications such as satellites, lighthouses, or in developing countries, batteries or additional power generators are often added as back-ups. Such stand-alone power systems permit operations at night and at other times of limited sunlight.

Concentrated Solar Power

Concentrated solar power (CSP), also called "concentrated solar thermal", uses lenses or mirrors and tracking systems to concentrate sunlight, then use the resulting heat to generate electricity from conventional steam-driven turbines.

A wide range of concentrating technologies exists: among the best known are the parabolic trough, the compact linear Fresnel reflector, the Stirling dish and the solar power tower. Various techniques are used to track the sun and focus light. In all of these systems a working fluid is heated by the concentrated sunlight, and is then used for power generation or energy storage. Thermal storage efficiently allows up to 24-hour electricity generation.

A parabolic trough consists of a linear parabolic reflector that concentrates light onto a receiver positioned along the reflector's focal line. The receiver is a tube positioned along the focal points of the linear parabolic mirror and is filled with a working fluid. The reflector is made to follow the sun during daylight hours by tracking along a single axis. Parabolic trough systems provide the best land-use factor of any solar technology. The SEGS plants in California and Acciona's Nevada Solar One near Boulder City, Nevada are representatives of this technology.

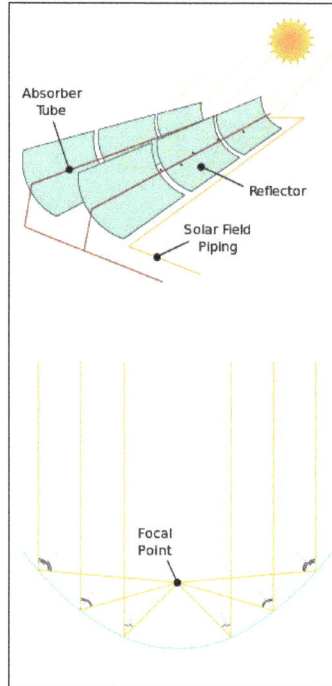

A parabolic collector concentrates sunlight onto a tube in its focal point.

Compact Linear Fresnel Reflectors are CSP-plants which use many thin mirror strips instead of parabolic mirrors to concentrate sunlight onto two tubes with working fluid. This has the advantage that flat mirrors can be used which are much cheaper than parabolic mirrors, and that more reflectors can be placed in the same amount of space, allowing more of the available sunlight to be used. Concentrating linear fresnel reflectors can be used in either large or more compact plants.

The *Stirling solar dish* combines a parabolic concentrating dish with a Stirling engine which normally drives an electric generator. The advantages of Stirling solar over photovoltaic cells are higher efficiency of converting sunlight into electricity and longer lifetime. Parabolic dish systems give the highest efficiency among CSP technologies. The 50 kW Big Dish in Canberra, Australia is an example of this technology.

A *solar power tower* uses an array of tracking reflectors (heliostats) to concentrate light on a central receiver atop a tower. Power towers can achieve higher (thermal-to-electricity conversion) efficiency than linear tracking CSP schemes and better energy storage capability than dish stirling technologies. The PS10 Solar Power Plant and PS20 solar power plant are examples of this technology.

Hybrid Systems

A hybrid system combines (C)PV and CSP with one another or with other forms of generation such as diesel, wind and biogas. The combined form of generation may enable the system to modulate power output as a function of demand or at least reduce the fluctuating nature of solar power and the consumption of non renewable fuel. Hybrid systems are most often found on islands.

- CPV/CSP system

  - A novel solar CPV/CSP hybrid system has been proposed, combining concentrator photovoltaics with the non-PV technology of concentrated solar power, or also known as concentrated solar thermal.

- ISCC system

  - The Hassi R'Mel power station in Algeria, is an example of combining CSP with a gas turbine, where a 25-megawatt CSP-parabolic trough array supplements a much larger 130 MW combined cycle gas turbine plant. Another example is the Yazd power station in Iran.

- PVT system

  - Hybrid PV/T, also known as *photovoltaic thermal hybrid solar collectors* convert solar radiation into thermal and electrical energy. Such a system combines a solar (PV) module with a solar thermal collector in a complementary way.

- CPVT system

  - A concentrated photovoltaic thermal hybrid (CPVT) system is similar to a PVT system. It uses concentrated photovoltaics (CPV) instead of conventional PV technology, and combines it with a solar thermal collector.

- PV diesel system

  - It combines a photovoltaic system with a diesel generator. Combinations with other renewables are possible and include wind turbines.

- PV-thermoelectric system

  - Thermoelectric, or "thermovoltaic" devices convert a temperature difference between dissimilar materials into an electric current. Solar cells use only the high frequency part of the radiation, while the low frequency heat energy is wasted. Several patents about the use of thermoelectric devices in tandem with solar cells have been filed.

The idea is to increase the efficiency of the combined solar/thermoelectric system to convert the solar radiation into useful electricity.

Deployment of Solar Power

Capacity in MW by Technology

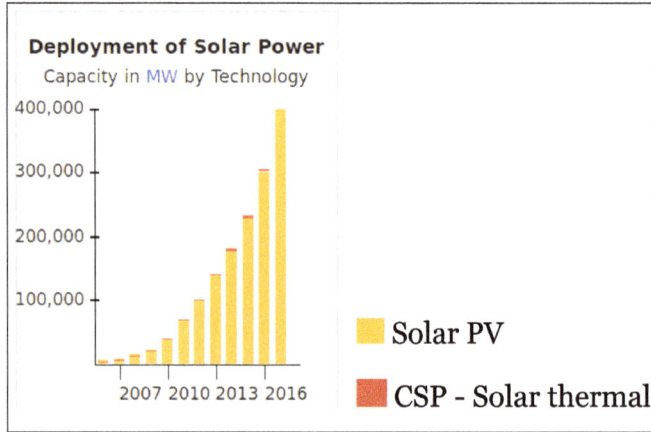

Worldwide deployment of solar power by technology since 2006

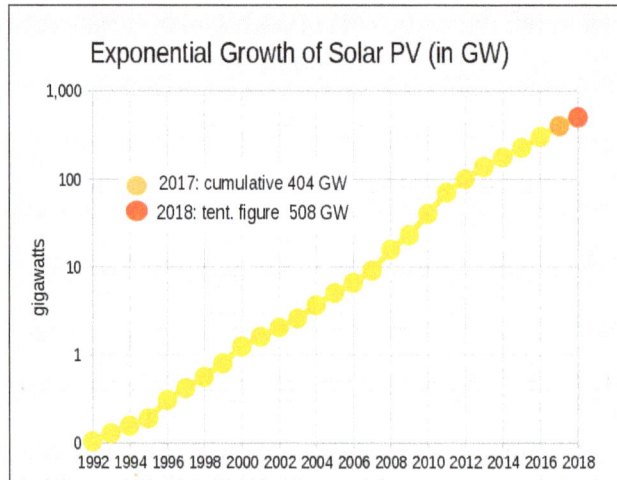

Solar Electricity Generation		
Year	Energy (TWh)	% of Total
2004	2.6	0.01%
2005	3.7	0.02%
2006	5.0	0.03%
2007	6.8	0.03%
2008	11.4	0.06%
2009	19.3	0.10%
2010	31.4	0.15%
2011	60.6	0.27%
2012	96.7	0.43%

2013	134.5	0.58%
2014	185.9	0.79%
2015	253.0	1.05%
2016	328.2	1.31%
2017	442.6	1.73%

The early development of solar technologies starting in the 1860s was driven by an expectation that coal would soon become scarce. Charles Fritts installed the world's first rooftop photovoltaic solar array, using 1%-efficient selenium cells, on a New York City roof in 1884. However, development of solar technologies stagnated in the early 20th century in the face of the increasing availability, economy, and utility of coal and petroleum. In 1974 it was estimated that only six private homes in all of North America were entirely heated or cooled by functional solar power systems. The 1973 oil embargo and 1979 energy crisis caused a reorganization of energy policies around the world and brought renewed attention to developing solar technologies. Deployment strategies focused on incentive programs such as the Federal Photovoltaic Utilization Program in the US and the Sunshine Program in Japan. Other efforts included the formation of research facilities in the United States (SERI, now NREL), Japan (NEDO), and Germany (Fraunhofer–ISE). Between 1970 and 1983 installations of photovoltaic systems grew rapidly, but falling oil prices in the early 1980s moderated the growth of photovoltaics from 1984 to 1996.

In the mid-1990s development of both, residential and commercial rooftop solar as well as utility-scale photovoltaic power stations began to accelerate again due to supply issues with oil and natural gas, global warming concerns, and the improving economic position of PV relative to other energy technologies. In the early 2000s, the adoption of feed-in tariffs—a policy mechanism, that gives renewables priority on the grid and defines a fixed price for the generated electricity—led to a high level of investment security and to a soaring number of PV deployments in Europe.

For several years, worldwide growth of solar PV was driven by European deployment, but has since shifted to Asia, especially China and Japan, and to a growing number of countries and regions all over the world, including, but not limited to, Australia, Canada, Chile, India, Israel, Mexico, South Africa, South Korea, Thailand, and the United States.

Worldwide growth of photovoltaics has averaged 40% per year from 2000 to 2013 and total installed capacity reached 303 GW at the end of 2016 with China having the most cumulative installations (78 GW) and Honduras having the highest theoretical percentage of annual electricity usage which could be generated by solar PV (12.5%). The largest manufacturers are located in China.

Concentrated solar power (CSP) also started to grow rapidly, increasing its capacity nearly tenfold from 2004 to 2013, albeit from a lower level and involving fewer countries than solar PV. As of the end of 2013, worldwide cumulative CSP-capacity reached 3,425 MW.

Forecasts

In 2010, the International Energy Agency predicted that global solar PV capacity could reach 3,000 GW or 11% of projected global electricity generation by 2050—enough to generate 4,500 TWh of electricity. Four years later, in 2014, the agency projected that, under its "high renewables" scenario, solar power could supply 27% of global electricity generation by 2050 (16% from PV and 11% from CSP).

Photovoltaic Power Stations

The Desert Sunlight Solar Farm is a 550 MW power plant in Riverside County, California, that uses thin-film CdTe-modules made by First Solar. As of November 2014, the 550 megawatt Topaz Solar Farm was the largest photovoltaic power plant in the world. This was surpassed by the 579 MW Solar Star complex. The current largest photovoltaic power station in the world is Longyangxia Dam Solar Park, in Gonghe County, Qinghai, China.

Concentrating Solar Power Stations

Ivanpah Solar Electric Generating System with all three towers under load during Clark Mountain Range seen in the distance.

Part of the 354 MW Solar Energy Generating Systems (SEGS) parabolic trough solar complex in northern San Bernardino County, California.

Commercial concentrating solar power (CSP) plants, also called "solar thermal power stations", were first developed in the 1980s. The 377 MW Ivanpah Solar Power Facility, located in California's Mojave Desert, is the world's largest solar thermal power plant project. Other large CSP plants include the Solnova Solar Power Station (150 MW), the Andasol solar power station (150 MW), and Extresol Solar Power Station (150 MW), all in Spain. The principal advantage of CSP is the ability to efficiently add thermal storage, allowing the dispatching of electricity over up to a 24-hour period. Since peak electricity demand typically occurs at about 5 pm, many CSP power plants use 3 to 5 hours of thermal storage.

Economics

Cost

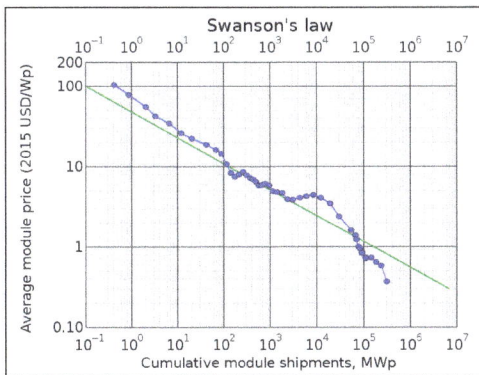

Swanson's law – the PV learning curve.

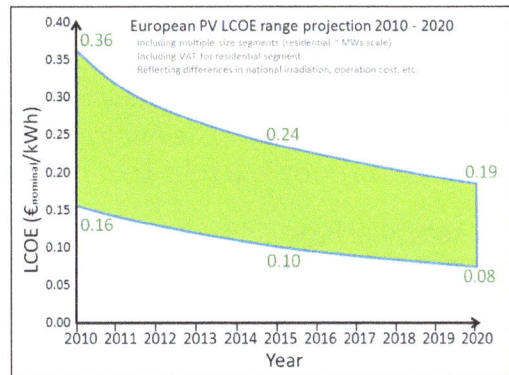

Solar PV – LCOE for Europe until 2020 (in euro-cts. per kWh).

The typical cost factors for solar power include the costs of the modules, the frame to hold them, wiring, inverters, labour cost, any land that might be required, the grid connection, maintenance and the solar insolation that location will receive. Adjusting for inflation, it cost $96 per watt for a solar module in the mid-1970s. Process improvements and a very large boost in production have brought that figure down to 68 cents per watt in February 2016, according to data from Bloomberg New Energy Finance. Palo Alto California signed a wholesale purchase agreement in 2016 that secured solar power for 3.7 cents per kilowatt-hour. And in sunny Dubai large-scale solar generated electricity sold in 2016 for just 2.99 cents per kilowatt-hour – "competitive with any form of fossil-based electricity — and cheaper than most."

Photovoltaic systems use no fuel, and modules typically last 25 to 40 years. Thus, capital costs make up most of the cost of solar power. Operations and maintenance costs for new utility-scale solar plants in the US are estimated to be 9 percent of the cost of photovoltaic electricity, and 17 percent of the cost of solar thermal electricity. Governments have created various financial incentives to encourage the use of solar power, such as feed-in tariff programs. Also, Renewable portfolio standards impose a government mandate that utilities generate or acquire a certain percentage of renewable power regardless of increased energy procurement costs. In most states,

RPS goals can be achieved by any combination of solar, wind, biomass, landfill gas, ocean, geothermal, municipal solid waste, hydroelectric, hydrogen, or fuel cell technologies.

Levelized Cost of Electricity

The PV industry has adopted levelized cost of electricity (LCOE) as the unit of cost. The electrical energy generated is sold in units of kilowatt-hours (kWh). As a rule of thumb, and depending on the local insolation, 1 watt-peak of installed solar PV capacity generates about 1 to 2 kWh of electricity per year. This corresponds to a capacity factor of around 10–20%. The product of the local cost of electricity and the insolation determines the break even point for solar power. The International Conference on Solar Photovoltaic Investments, organized by EPIA, has estimated that PV systems will pay back their investors in 8 to 12 years. As a result, since 2006 it has been economical for investors to install photovoltaics for free in return for a long term power purchase agreement. Fifty percent of commercial systems in the United States were installed in this manner in 2007 and over 90% by 2009.

Shi Zhengrong has said that, as of 2012, unsubsidised solar power is already competitive with fossil fuels in India, Hawaii, Italy and Spain. He said "We are at a tipping point. No longer are renewable power sources like solar and wind a luxury of the rich. They are now starting to compete in the real world without subsidies". "Solar power will be able to compete without subsidies against conventional power sources in half the world by 2015".

Current Installation Prices

Utility-scale PV system prices		
Country	Cost ($/W)	Year
Australia	2.0	2013
China	1.4	2013
France	2.2	2013
Germany	1.4	2013
Italy	1.5	2013
Japan	2.9	2013
United Kingdom	1.9	2013
United States	1.25	June 2016

In its 2014 edition of the Technology Roadmap: Solar Photovoltaic Energy report, the International Energy Agency (IEA) published prices for residential, commercial and utility-scale PV systems for eight major markets as of 2013. However, DOE's SunShot Initiative has reported much lower U.S. installation prices. In 2014, prices continued to decline. The SunShot Initiative modeled U.S. system prices to be in the range of $1.80 to $3.29 per watt. Other sources identify similar price ranges of $1.70 to $3.50 for the different market segments in the U.S. and in the highly penetrated German market, prices for residential and small commercial rooftop systems of up to 100 kW declined to $1.36 per watt (€1.24/W) by the end of 2014. In 2015, Deutsche Bank estimated costs for small residential rooftop systems in the U.S. around $2.90 per watt. Costs for utility-scale systems in China and India were estimated as low as $1.00 per watt.

Grid Parity

Grid parity, the point at which the cost of photovoltaic electricity is equal to or cheaper than the price of grid power, is more easily achieved in areas with abundant sun and high costs for electricity such as in California and Japan. In 2008, the levelized cost of electricity for solar PV was $0.25/kWh or less in most of the OECD countries. By late 2011, the fully loaded cost was predicted to fall below $0.15/kWh for most of the OECD and to reach $0.10/kWh in sunnier regions. These cost levels are driving three emerging trends: vertical integration of the supply chain, origination of power purchase agreements (PPAs) by solar power companies, and unexpected risk for traditional power generation companies, grid operators and wind turbine manufacturers.

Grid parity was first reached in Spain in 2013, Hawaii and other islands that otherwise use fossil fuel (diesel fuel) to produce electricity, and most of the US is expected to reach grid parity by 2015.

In 2007, General Electric's Chief Engineer predicted grid parity without subsidies in sunny parts of the United States by around 2015; other companies predicted an earlier date: the cost of solar power will be below grid parity for more than half of residential customers and 10% of commercial customers in the OECD, as long as grid electricity prices do not decrease through 2010.

Productivity by Location

The productivity of solar power in a region depends on solar irradiance, which varies through the day and is influenced by latitude and climate.

The locations with highest annual solar irradiance lie in the arid tropics and subtropics. Deserts lying in low latitudes usually have few clouds, and can receive sunshine for more than ten hours a day. These hot deserts form the *Global Sun Belt* circling the world. This belt consists of extensive swathes of land in Northern Africa, Southern Africa, Southwest Asia, Middle East, and Australia, as well as the much smaller deserts of North and South America. Africa's eastern Sahara Desert, also known as the Libyan Desert, has been observed to be the sunniest place on Earth according to NASA.

Different measurements of solar irradiance (direct normal irradiance, global horizontal irradiance) are mapped below :

North America

South America

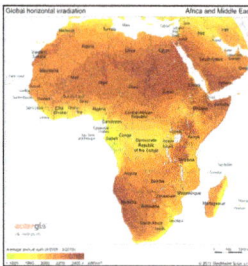
Africa and Middle East

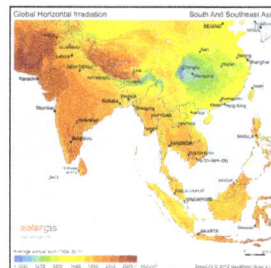
South and South-East Asia

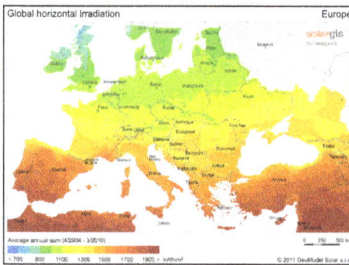
Europe

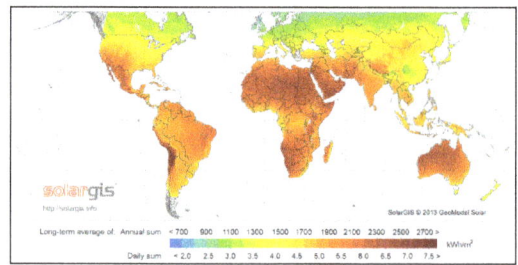
World

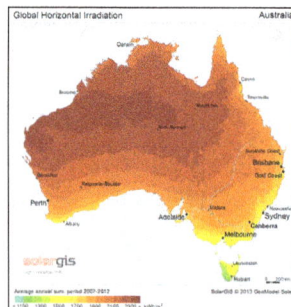
Australia

Self Consumption

In cases of self consumption of the solar energy, the payback time is calculated based on how much electricity is not purchased from the grid. For example, in Germany, with electricity prices of 0.25 €/kWh and insolation of 900 kWh/kW, one kWp will save €225 per year, and with an installation cost of 1700 €/KWp the system cost will be returned in less than seven years. However, in many cases, the patterns of generation and consumption do not coincide, and some or all of the energy is fed back into the grid. The electricity is sold, and at other times when energy is taken from the grid, electricity is bought. The relative costs and prices obtained affect the economics. In many markets, the price paid for sold PV electricity is significantly lower than the price of bought electricity, which incentivizes self consumption. Moreover, separate self consumption incentives have been used in e.g. Germany and Italy. Grid interaction regulation has also included limitations of grid feed-in in some regions in Germany with high amounts of installed PV capacity. By increasing self consumption, the grid feed-in can be limited without curtailment, which wastes electricity.

A good match between generation and consumption is key for high self consumption, and should be considered when deciding where to install solar power and how to dimension the installation. The match can be improved with batteries or controllable electricity consumption. However, batteries are expensive and profitability may require provision of other services from them besides self consumption increase. Hot water storage tanks with electric heating with heat pumps or resistance heaters can provide low-cost storage for self consumption of solar power. Shiftable loads, such as dishwashers, tumble dryers and washing machines, can provide controllable consumption with only a limited effect on the users, but their effect on self consumption of solar power may be limited.

Energy Pricing and Incentives

The political purpose of incentive policies for PV is to facilitate an initial small-scale deployment to begin to grow the industry, even where the cost of PV is significantly above grid parity, to allow the industry to achieve the economies of scale necessary to reach grid parity. The policies are implemented to promote national energy independence, high tech job creation and reduction of CO_2 emissions. Three incentive mechanisms are often used in combination as investment subsidies: the authorities refund part of the cost of installation of the system, the electricity utility buys PV electricity from the producer under a multiyear contract at a guaranteed rate, and Solar Renewable Energy Certificates (SRECs).

Rebates

With investment subsidies, the financial burden falls upon the taxpayer, while with feed-in tariffs the extra cost is distributed across the utilities' customer bases. While the investment subsidy may be simpler to administer, the main argument in favour of feed-in tariffs is the encouragement of quality. Investment subsidies are paid out as a function of the nameplate capacity of the installed system and are independent of its actual power yield over time, thus rewarding the overstatement of power and tolerating poor durability and maintenance. Some electric companies offer rebates to their customers, such as Austin Energy in Texas, which offers $2.50/watt installed up to $15,000.

Net Metering

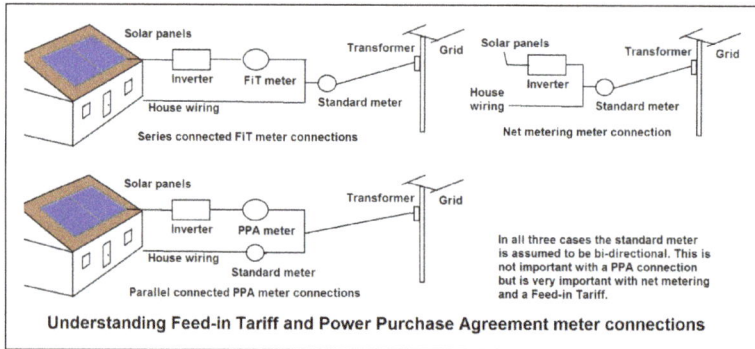

Understanding Feed-in Tariff and Power Purchase Agreement meter connections

Net metering, unlike a feed-in tariff, requires only one meter, but it must be bi-directional.

In net metering the price of the electricity produced is the same as the price supplied to the consumer, and the consumer is billed on the difference between production and consumption. Net metering can usually be done with no changes to standard electricity meters, which accurately measure power in both directions and automatically report the difference, and because it allows homeowners and businesses to generate electricity at a different time from consumption, effectively using the grid as a giant storage battery. With net metering, deficits are billed each month while surpluses are rolled over to the following month. Best practices call for perpetual roll over of kWh credits. Excess credits upon termination of service are either lost, or paid for at a rate ranging from wholesale to retail rate or above, as can be excess annual credits. In New Jersey, annual excess credits are paid at the wholesale rate, as are left over credits when a customer terminates service.

Feed-in Tariffs

With feed-in tariffs, the financial burden falls upon the consumer. They reward the number of kilowatt-hours produced over a long period of time, but because the rate is set by the authorities, it may result in perceived overpayment. The price paid per kilowatt-hour under a feed-in tariff exceeds the price of grid electricity. Net metering refers to the case where the price paid by the utility is the same as the price charged.

The complexity of approvals in California, Spain and Italy has prevented comparable growth to Germany even though the return on investment is better. In some countries, additional incentives are offered for building-integrated photovoltaics (BIPV) compared to stand alone PV.

- France + EUR 0.16 /kWh (compared to semi-integrated) or + EUR 0.27/kWh (compared to stand alone)

- Italy + EUR 0.04–0.09 kWh

- Germany + EUR 0.05/kWh (facades only)

Solar Renewable Energy Credits

Alternatively, Solar Renewable Energy Certificates (SRECs) allow for a market mechanism to set the price of the solar generated electricity subsidy. In this mechanism, a renewable energy production or consumption target is set, and the utility (more technically the Load Serving Entity) is obliged to purchase renewable energy or face a fine (Alternative Compliance Payment or ACP). The producer is credited for an SREC for every 1,000 kWh of electricity produced. If the utility buys this SREC and retires it, they avoid paying the ACP. In principle this system delivers the cheapest renewable energy, since the all solar facilities are eligible and can be installed in the most economic locations. Uncertainties about the future value of SRECs have led to long-term SREC contract markets to give clarity to their prices and allow solar developers to pre-sell and hedge their credits.

Financial incentives for photovoltaics differ across countries, including Australia, China, Germany, Israel, Japan, and the United States and even across states within the US.

The Japanese government through its Ministry of International Trade and Industry ran a successful programme of subsidies from 1994 to 2003. By the end of 2004, Japan led the world in installed PV capacity with over 1.1 GW.

In 2004, the German government introduced the first large-scale feed-in tariff system, under the German Renewable Energy Act, which resulted in explosive growth of PV installations in Germany. At the outset the FIT was over 3x the retail price or 8x the industrial price. The principle behind the German system is a 20-year flat rate contract. The value of new contracts is programmed to decrease each year, in order to encourage the industry to pass on lower costs to the end users. The programme has been more successful than expected with over 1GW installed in 2006, and political pressure is mounting to decrease the tariff to lessen the future burden on consumers.

Subsequently, Spain, Italy, Greece—that enjoyed an early success with domestic solar-thermal installations for hot water needs—and France introduced feed-in tariffs. None have replicated the programmed decrease of FIT in new contracts though, making the German incentive relatively less and less attractive compared to other countries. The French and Greek FIT offer a high premium (EUR 0.55/kWh) for building integrated systems. California, Greece, France and Italy have 30–50% more insolation than Germany making them financially more attractive. The Greek domestic "solar roof" programme (adopted in June 2009 for installations up to 10 kW) has internal rates of return of 10–15% at current commercial installation costs, which, furthermore, is tax free.

In 2006 California approved the 'California Solar Initiative', offering a choice of investment subsidies or FIT for small and medium systems and a FIT for large systems. The small-system FIT of $0.39 per kWh (far less than EU countries) expires in just

5 years, and the alternate "EPBB" residential investment incentive is modest, averaging perhaps 20% of cost. All California incentives are scheduled to decrease in the future depending as a function of the amount of PV capacity installed.

At the end of 2006, the Ontario Power Authority (OPA, Canada) began its Standard Offer Program, a precursor to the Green Energy Act, and the first in North America for distributed renewable projects of less than 10 MW. The feed-in tariff guaranteed a fixed price of $0.42 CDN per kWh over a period of twenty years. Unlike net metering, all the electricity produced was sold to the OPA at the given rate.

Grid Integration

In the above figure construction of the Salt Tanks which provide efficient thermal energy storage so that output can be provided after sunset, and output can be scheduled to meet demand requirements. The 280 MW Solana Generating Station is designed to provide six hours of energy storage. This allows the plant to generate about 38% of its rated capacity over the course of a year.

Pumped-storage hydroelectricity (PSH). This facility in Geesthacht, Germany, also includes a solar array.

The overwhelming majority of electricity produced worldwide is used immediately, since storage is usually more expensive and because traditional generators can adapt to demand. Both solar power and wind power are variable renewable energy, meaning that all available output must be taken whenever it is available by moving through transmission lines to where it can be used now. Since solar energy is not available at night, storing its energy is potentially an important issue particularly in off-grid and for future 100% renewable energy scenarios to have continuous electricity availability.

Solar electricity is inherently variable and predictable by time of day, location, and seasons. In addition solar is intermittent due to day/night cycles and unpredictable weather. How much of a special challenge solar power is in any given electric utility varies significantly. In a summer peak utility, solar is well matched to daytime cooling demands. In winter peak utilities, solar displaces other forms of generation, reducing their capacity factors.

In an electricity system without grid energy storage, generation from stored fuels (coal, biomass, natural gas, nuclear) must go up and down in reaction to the rise and fall of solar electricity. While hydroelectric and natural gas plants can quickly respond to changes in load, coal, biomass and nuclear plants usually take considerable time to respond to load and can only be scheduled to follow the predictable variation. Depending on local circumstances, beyond about 20–40% of total generation, grid-connected intermittent sources like solar tend to require investment in some combination of grid interconnections, energy storage or demand side management. Integrating large amounts of solar power with existing generation equipment has caused issues in some cases. For example, in Germany, California and Hawaii, electricity prices have been known to go negative when solar is generating a lot of power, displacing existing baseload generation contracts.

Conventional hydroelectricity works very well in conjunction with solar power; water can be held back or released from a reservoir as required. Where a suitable river is not available, pumped-storage hydroelectricity uses solar power to pump water to a high reservoir on sunny days, then the energy is recovered at night and in bad weather by releasing water via a hydroelectric plant to a low reservoir where the cycle can begin again. This cycle can lose 20% of the energy to round trip inefficiencies, this plus the construction costs add to the expense of implementing high levels of solar power.

Concentrated solar power plants may use thermal storage to store solar energy, such as in high-temperature molten salts. These salts are an effective storage medium because they are low-cost, have a high specific heat capacity, and can deliver heat at temperatures compatible with conventional power systems. This method of energy storage is used, for example, by the Solar Two power station, allowing it to store 1.44 TJ in its 68 m^3 storage tank, enough to provide full output for close to 39 hours, with an efficiency of about 99%.

In stand alone PV systems batteries are traditionally used to store excess electricity. With grid-connected photovoltaic power system, excess electricity can be sent to the electrical grid. Net metering and feed-in tariff programs give these systems a credit for the electricity they produce. This credit offsets electricity provided from the grid when the system cannot meet demand, effectively trading with the grid instead of storing excess electricity. Credits are normally rolled over from month to month and any remaining surplus settled annually. When wind and solar are a small fraction of the grid power, other generation techniques can adjust their output appropriately, but as these forms of variable power grow, additional balance on the grid is needed. As prices are rapidly declining, PV systems increasingly use rechargeable batteries to store a surplus to be later used at night. Batteries used for grid-storage stabilize the electrical grid by leveling out peak loads usually for several minutes, and in rare cases for hours. In the future, less expensive batteries could play an important role on the electrical grid, as they can charge during periods when generation exceeds demand and feed their stored energy into the grid when demand is higher than generation.

Although not permitted under the US National Electric Code, it is technically possible to have a "plug and play" PV microinverter.System design would enable such systems to meet all technical, though not all safety requirements. There are several companies selling plug and play solar systems available on the web, but there is a concern that if people install their own it will reduce the enormous employment advantage solar has over fossil fuels.

Common battery technologies used in today's home PV systems include, the valve regulated lead-acid battery– a modified version of the conventional lead–acid battery, nickel–cadmium and lithium-ion batteries. Lead-acid batteries are currently the predominant technology used in small-scale, residential PV systems, due to their high reliability, low self discharge and investment and maintenance costs, despite shorter lifetime and lower energy density. Lithium-ion batteries have the potential to replace lead-acid batteries in the near future, as they are being intensively developed and lower prices are expected due to economies of scale provided by large production facilities such as the Gigafactory. In addition, the Li-ion batteries of plug-in electric cars may serve as a future storage devices in a vehicle-to-grid system. Since most vehicles are parked an average of 95% of the time, their batteries could be used to let electricity flow from the car to the power lines and back. Other rechargeable batteries used for distributed PV systems include, sodium–sulfur and vanadium redox batteries, two prominent types of a molten salt and a flow battery, respectively.

The combination of wind and solar PV has the advantage that the two sources complement each other because the peak operating times for each system occur at different times of the day and year. The power generation of such solar hybrid power systems is therefore more constant and fluctuates less than each of the two component subsystems. Solar power is seasonal, particularly in northern/southern climates, away from the equator, suggesting a need for long term seasonal storage in a medium such as hydrogen or pumped hydroelectric. The Institute for Solar Energy Supply Technology of the University of Kassel pilot-tested a combined power plant linking solar, wind, biogas and pumped-storage hydroelectricity to provide load-following power from renewable sources.

Research is also undertaken in this field of artificial photosynthesis. It involves the use of nanotechnology to store solar electromagnetic energy in chemical bonds, by splitting water to produce hydrogen fuel or then combining with carbon dioxide to make bio-polymers such as methanol. Many large national and regional research projects on artificial photosynthesis are now trying to develop techniques integrating improved light capture, quantum coherence methods of electron transfer and cheap catalytic materials that operate under a variety of atmospheric conditions. Senior researchers in the field have made the public policy case for a Global Project on Artificial Photosynthesis to address critical energy security and environmental sustainability issues.

Environmental Impacts

Part of the Senftenberg Solarpark, a solar photovoltaic power plant located on former open-pit mining areas close to the city of Senftenberg, in Eastern Germany. The 78 MW Phase 1 of the plant was completed within three months.

Unlike fossil fuel based technologies, solar power does not lead to any harmful emissions during operation, but the production of the panels leads to some amount of pollution.

Greenhouse Gases

The life-cycle greenhouse-gas emissions of solar power are in the range of 22 to 46 gram (g) per kilowatt-hour (kWh) depending on if solar thermal or solar PV is being analyzed, respectively. With this potentially being decreased to 15 g/kWh in the future. For comparison (of weighted averages), a combined cycle gas-fired power plant emits some 400–599 g/kWh, an oil-fired power plant 893 g/kWh, a coal-fired power plant 915–994 g/kWh or with carbon capture and storage some 200 g/kWh, and a geothermal high-temp. power plant 91–122 g/kWh. The life cycle emission intensity of hydro, wind and nuclear power are lower than solar's as of 2011 Life-cycle greenhouse-gas emissions of energy sources. Similar to all energy sources were their total life cycle emissions primarily lay in the construction and transportation phase, the switch to low carbon power in the manufacturing and transportation of solar devices would further reduce carbon emissions. BP Solar owns two factories built by Solarex (one in Maryland, the other in Virginia) in which all of the energy used to manufacture solar panels is produced by solar panels. A 1-kilowatt system eliminates the burning of approximately 170 pounds of coal, 300 pounds of carbon dioxide from being released into the atmosphere, and saves up to 105 gallons of water consumption monthly.

The US National Renewable Energy Laboratory (NREL), in harmonizing the disparate estimates of life-cycle GHG emissions for solar PV, found that the most critical parameter was the solar insolation of the site: GHG emissions factors for PV solar are inversely proportional to insolation. For a site with insolation of 1700 kWh/m2/year, typical of southern Europe, NREL researchers estimated GHG emissions of 45 gCO$_2$e/kWh. Using the same assumptions, at Phoenix, USA, with insolation of 2400 kWh/m2/year, the GHG emissions factor would be reduced to 32 g of CO$_2$e/kWh.

The New Zealand Parliamentary Commissioner for the Environment found that the solar PV would have little impact on the country's greenhouse gas emissions. The country already generates 80 percent of its electricity from renewable resources (primarily hydroelectricity and geothermal) and national electricity usage peaks on winter evenings whereas solar generation peaks on summer afternoons, meaning a large uptake of solar PV would end up displacing other renewable generators before fossil-fueled power plants.

Energy Payback

The energy payback time (EPBT) of a power generating system is the time required to generate as much energy as is consumed during production and lifetime operation of the system. Due to improving production technologies the payback time has been decreasing constantly since the introduction of PV systems in the energy market. In 2000 the energy payback time of PV systems was estimated as 8 to 11 years and in 2006 this was estimated to be 1.5 to 3.5 years for crystalline silicon PV systems and 1–1.5 years for thin film technologies (S. Europe). These figures fell to 0.75–3.5 years in 2013, with an average of about 2 years for crystalline silicon PV and CIS systems.

Another economic measure, closely related to the energy payback time, is the energy returned on energy invested (EROEI) or energy return on investment (EROI), which is the ratio of electricity generated divided by the energy required to build *and maintain* the equipment. (This is not the same as the economic return on investment (ROI), which varies according to local energy prices, subsidies available and metering techniques.) With expected lifetimes of 30 years, the EROEI of PV systems are in the range of 10 to 30, thus generating enough energy over their lifetimes to reproduce themselves many times (6–31 reproductions) depending on what type of material, balance of system (BOS), and the geographic location of the system.

Water Use

Solar power includes plants with among the lowest water consumption per unit of electricity (photovoltaic), and also power plants with among the highest water consumption (concentrating solar power with wet-cooling systems).

Photovoltaic power plants use very little water for operations. Life-cycle water consumption for utility-scale operations is estimated to be 12 gallons per mega-watt-hour for flat-panel PV solar. Only wind power, which consumes essentially no water during operations, has a lower water consumption intensity.

Concentrating solar power plants with wet-cooling systems, on the other hand, have the highest water-consumption intensities of any conventional type of electric power plant; only fossil-fuel plants with carbon-capture and storage may have higher water intensities. A 2013 study comparing various sources of electricity found that the median water consumption during operations of concentrating solar power plants with wet cooling was 810 ga/MWhr for power tower plants and 890 gal/MWhr for trough plants. This was higher than the operational water consumption (with cooling towers) for nuclear (720 gal/MWhr), coal (530 gal/MWhr), or natural gas (210). A 2011 study by the National Renewable Energy Laboratory came to similar conclusions: for power plants with cooling towers, water consumption during operations was 865 gal/MWhr for CSP trough, 786 gal/MWhr for CSP tower, 687 gal/MWhr for coal, 672 gal/MWhr for nuclear, and 198 gal/MWhr for natural gas. The Solar Energy Industries Association noted that the Nevada Solar One trough CSP plant consumes 850 gal/MWhr. The issue of water consumption is heightened because CSP plants are often located in arid environments where water is scarce.

In 2007, the US Congress directed the Department of Energy to report on ways to reduce water consumption by CSP. The subsequent report noted that dry cooling technology was available that, although more expensive to build and operate, could reduce water consumption by CSP by 91 to 95 percent. A hybrid wet/dry cooling system could reduce water consumption by 32 to 58 percent. A 2015 report by NREL noted that of the 24 operating CSP power plants in the US, 4 used dry cooling systems. The four dry-cooled systems were the three power plants at the Ivanpah Solar Power Facility near Barstow, California, and the Genesis Solar Energy Project in Riverside County, California. Of 15 CSP projects under construction or development in the US as of March 2015, 6 were wet systems, 7 were dry systems, 1 hybrid, and 1 unspecified.

Although many older thermoelectric power plants with once-through cooling or cooling ponds *use* more water than CSP, meaning that more water passes through their systems, most of the cooling water returns to the water body available for other uses, and they *consume* less water by evaporation. For instance, the median coal power plant in the US with once-through cooling uses 36,350 gal/MWhr, but only 250 gal/MWhr (less than one percent) is lost through evaporation. Since the 1970s, the majority of US power plants have used recirculating systems such as cooling towers rather than once-through systems.

Other Issues

One issue that has often raised concerns is the use of cadmium (Cd), a toxic heavy metal that has the tendency to accumulate in ecological food chains. It is used as semiconductor component in CdTe solar cells and as a buffer layer for certain CIGS cells in the form of cadmium sulfide. The amount of cadmium used in thin-film solar cells is relatively small (5–10 g/m²) and with proper recycling and emission control techniques in place the cadmium emissions from module production can be almost zero. Current PV technologies lead to cadmium emissions of 0.3–0.9 microgram/kWh over the whole life-cycle. Most of these emissions arise through the use of coal power for the manufacturing of the modules, and coal and lignite combustion leads to much higher emissions of cadmium. Life-cycle cadmium emissions from coal is 3.1 microgram/kWh, lignite 6.2, and natural gas 0.2 microgram/kWh.

In a life-cycle analysis it has been noted, that if electricity produced by photovoltaic panels were used to manufacture the modules instead of electricity from burning coal, cadmium emissions from coal power usage in the manufacturing process could be entirely eliminated.

In the case of crystalline silicon modules, the solder material, that joins together the copper strings of the cells, contains about 36 percent of lead (Pb). Moreover, the paste used for screen printing front and back contacts contains traces of Pb and sometimes Cd as well. It is estimated that about 1,000 metric tonnes of Pb have been used for 100 gigawatts of c-Si solar modules. However, there is no fundamental need for lead in the solder alloy.

Some media sources have reported that concentrated solar power plants have injured or killed large numbers of birds due to intense heat from the concentrated sunrays. This adverse effect does not apply to PV solar power plants, and some of the claims may have been overstated or exaggerated.

A 2014-published life-cycle analysis of land use for various sources of electricity concluded that the large-scale implementation of solar and wind potentially reduces pollution-related environmental impacts. The study found that the land-use footprint, given in square meter-years per megawatt-hour (m²a/MWh), was lowest for wind, natural gas and rooftop PV, with 0.26, 0.49 and 0.59, respectively, and followed by utility-scale solar PV with 7.9. For CSP, the footprint was 9 and 14, using parabolic troughs and solar towers, respectively. The largest footprint had coal-fired power plants with 18 m²a/MWh.

Emerging Technologies

Concentrator Photovoltaics

Concentrator photovoltaics (CPV) systems employ sunlight concentrated onto photovoltaic surfaces for the purpose of electrical power production. Contrary to conventional photovoltaic systems, it uses lenses and curved mirrors to focus sunlight onto small, but highly efficient, multi-junction solar cells. Solar concentrators of all varieties may be used, and these are often mounted on a solar tracker in order to keep the focal point upon the cell as the sun moves across the sky. Luminescent solar concentrators (when combined with a PV-solar cell) can also be regarded as a CPV system. Concentrated photovoltaics are useful as they can improve efficiency of PV-solar panels drastically.

CPV modules on dual axis solar trackers in Golmud, China.

In addition, most solar panels on spacecraft are also made of high efficient multi-junction photovoltaic cells to derive electricity from sunlight when operating in the inner Solar System.

Floatovoltaics

Floatovoltaics are an emerging form of PV systems that float on the surface of irrigation canals, water reservoirs, quarry lakes, and tailing ponds. Several systems exist in France, India, Japan, Korea, the United Kingdom and the United States. These systems reduce the need of valuable land area, save drinking water that would otherwise be lost through evaporation, and show a higher efficiency of solar energy conversion, as the panels are kept at a cooler temperature than they would be on land. Although not floating, other dual-use facilities with solar power include fisheries.

Solar Power Forecasting

Solar power forecasting involves knowledge of the Sun´s path, the atmosphere's condition, the scattering processes and the characteristics of a solar energy plant which utilizes the Sun's energy to create solar power. Solar photovoltaic systems transform solar energy into electric power. The power output depends on the incoming radiation and on the solar panel characteristics. Photovoltaic power production is increasing nowadays. Forecast information is essential for an efficient use, the management of the electricity grid and for solar energy trading. Common solar forecasting method include stochastic learning method, local and remote sensing method, and hybrid method.

Generation Forecasting

The energy generation forecasting problem is closely linked to the problem of *weather variables forecasting*. Indeed, this problem is usually split into two parts, on one hand focusing on the forecasting of solar PV or any other meteorological variable and on the other hand estimating the amount of energy that a concrete power plant will produce with the estimated meteorological resource. In general, the way to deal with this difficult problem is usually related to the spatial and temporal scales we are interested in, which yields to different approaches that can be found in the literature. In this sense, it is useful to classify these techniques depending on the forecasting horizon, so it is possible to distinguish between *now-casting* (forecasting 3–4 hours ahead), *short-term forecasting* (up to 7 days ahead) and *long-term forecasting* (months, years) Solar radiation closely follows the physical and biological development of the earth. Its spatial and sequential heterogeneity powerfully influence the forcing of environmental and hydrological organisms by manipulating air temperature, soil moisture and vapor transpiration, snow cover and lots of photochemical procedures. Therefore, solar radiation drives place efficiency and plant life allotment, organism a key feature in undeveloped and forestry sciences that be obliged to be known precisely.

The quantity of solar radiation obtainable at the earth' surface is at the outset controlled at worldwide balance, organism above all precious by the Sun Earth geometry and the atmosphere. On the other hand, a complete explanation of its freedom time unpredictability require deliberation of limited procedure which frequently turn out to be also applicable, as is the casing in mountainous region. Predominantly, limited territory adjust the inward bound solar radiation by shadow casts, slope of elevation, surface gradient and compass reading, as a result, precise spatial model of inward bound solar radiation be supposed to regard as the pressure of the terrain surface. In the final time, more than a few events to consist of the confined terrain special effects in the solar radiation countryside have been projected, such as the use of Geographical Information Systems (GIS), artificial intelligence or post dispensation of satellite stand technique. Solar radiation can be also evaluated using numerical weather forecast (NWP) models. Nevertheless, the space and time balance determined with them and the incomplete computational ability frequently avoid the deliberation of terrain connected property.

Otherwise, exclamation technique agree to us to acquire spatially persistent database from data evidence at inaccessible station greater than wide region. Even though their dependability is powerfully needy on the opening coldness between position, they eventually rely on experiential statistics, which have a superior precision than extra method. Therefore, while an adequate footage spatial thickness is accessible, disturbance method are preferred. Conventionally, solar radiation has not been as densely example as additional variables as temperature or rainfall, therefore the ease of use of capacity is frequently in short supply. Though, the number of experimental system which record solar radiation has developed and interruption has been converted into an appropriate technique for solar radiation evaluation. Nevertheless, radiometric stations are frequently come together approximately farmland or occupied region, typically during basin and plane area, while mountains at rest require enough footage thickness. This truth is particularly applicable afford the tall spatial unpredictability of solar radiation in these province. As an outcome, particular interruption method that tolerate include outdoor foundation should be used to make clear this extra spatial unpredictability.

Several diverse spatial interruption techniques can be established. On the other hand, data ease of use in mountainous region is often extremely restricted. As a result, it is hard to construct a precise solar radiation climatology in hilly area to be used in environmental science, climate change.

Solar radiation is a hardly illustration changeable with reverence to supplementary ecological variables such as temperature or precipitation, in fraction payable to the high maintenance price of the necessary radiometric sensors. It is extremely sensitive to ecological feature on or after local to limited balance. Predominantly, terrain surface confronts the conventional interruption method while forecast through far above the ground spatial decision are required, particularly due to the lack of measurement stations in mountainous areas. Geo-statistics front a stochastic move toward to resolve the spatial forecast difficulty that stop dependence on before imagine deterministic models and permit us to consist of the consequence of outside in sequence foundation stand on investigation data-sets.

Nowcasting

The term "Nowcasting" in the context of solar power forecasting, generally refers to same spatial and temporal scales as meteorological Nowcasting, which focuses on forecast horizons from a few minutes ahead, out to 4–6 hours ahead. In general, the 'Nowcasting' forecast horizon refers to those not well-served by global numerical weather prediction models, which produce outputs at up to hourly resolutions and only update every 6 hours. Solar power nowcasting then refers to the prediction of solar power output (or energy generation) over time horizons of tens to hundreds of minutes ahead of time with up to 90% predictability. It has historically been very important for electrical grid operators in order to guarantee the matching of supply and demand on energy markets. Such solar power nowcasting services are usually related to temporal resolutions of 5 to 15 minutes, with updates as frequent as every 5 minutes. The regular updates and relatively high resolutions required from these methods require automatic weather data acquisition and processing techniques, which are chiefly accomplished by two primary means:

1. Statistical techniques: These are usually based on time series processing of measurement data, including meteorological observations and power output measurements from a solar power facility. What then follows is the creation of a training dataset to tune the parameters of a model, before evaluation of model performance against a separate testing dataset. This class of techniques includes the use of any kind of statistical approach, such as autoregressive moving averages (ARMA, ARIMA, etc.), as well as machine learning techniques such as neural networks, support vector machines (etc.). These approaches are usually benchmarked to a persistence approach in order to evaluate their improvements. This persistence approach just assumes that any variable at time step t is the value it took in a previous time.

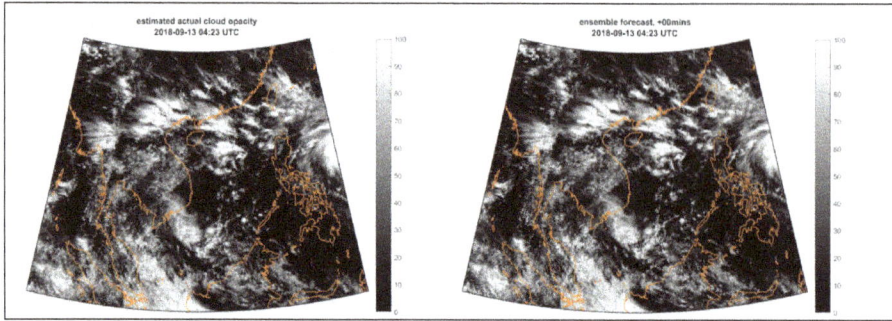

An example of satellite-based cloud cover nowcasting, which is used
to generate predication of solar power outputs.

2. Satellite based methods: These methods leverage the several geostationary
    Earth observing weather satellites (such as Meteosat Second Generation (MSG)
    fleet) to detect, characterise, track and predict the future locations of cloud
    cover. These satellites make it possible to generate solar power forecasts over
    broad regions through the application of image processing and forecasting al-
    gorithms. Key forecasting algorithms include cloud motion vectors (CMVs).
    Relevant methods for applying physical models based on satellite image pro-
    cessing techniques provide an estimation of future atmospheric values can be
    found in *Alvarez et al.* 2010.

Short-Term Solar Power Forecasting

Short-term forecasting provides predictions up to 7 days ahead. This kind of forecast is
also valuable for grid operators in order to make decisions of grid operation, as well as,
for electric market operators. Under this perspective, the meteorological resources are
estimated at a different temporal and spatial resolution. This implies that meteorological variables and phenomena are looked from a more general perspective, not as local as
nowcasting services do. In this sense, most of the approaches make use of different numerical weather prediction models (NWP) that provide an initial estimation of weather variables. Currently, several models are available for this purpose, such as Global
Forecast System (GFS) or data provided by the European Center for Medium Range
Weather Forecasting (ECMWF). These two models are considered the state of the art
of global forecast models, which provide meteorological forecasts all over the world. In
order to increase spatial and temporal resolution of these models, other models have
been developed which are generally called mesoscale models. Among others, HIRLAM,
WRF or MM5 are the most representative of these models since they are widely used
by different communities. To run these models a wide expertise is needed in order to
obtain accurate results, due to the wide variety of parameters that can be configured in
the models. In addition, sophisticated techniques such as data assimilation might be
used in order to produce more realistic simulations. Finally, some communities argue
for the use of post-processing techniques, once the models' output is obtained, in order
to obtain a probabilistic point of view of the accuracy of the output. This is usually done

with ensemble techniques that mix different outputs of different models perturbed in strategic meteorological values and finally provide a better estimate of those variables and a degree of uncertainty, like in the model proposed by Bacher et al.

Long-Term Solar Power Forecasting

Long-term forecasting usually refers to forecasting of the annual or monthly available resource. This is useful for energy producers and to negotiate contracts with financial entities or utilities that distribute the generated energy. In general, these long-term forecasting is usually done at a lower scale than any of the two previous approaches. Hence, most of these models are run with mesoscale models fed with reanalysis data as input and whose output is postprocessed with statistical approaches based on measured data.

Energetic Models

Any output from any model described above must then be converted to the electric energy that a particular solar PV plant will produce. This step is usually done with statistical approaches that try to correlate the amount of available resource with the metered power output. The main advantage of these methods is that the meteorological prediction error, which is the main component of the global error, might be reduced taking into account the uncertainty of the prediction. As it was mentioned before and detailed in *Heinemann et al.* these statistical approaches comprises from ARMA models, neural networks, support vector machines, etc. On the other hand, there also exist theoretical models that describe how a power plant converts the meteorological resource into electric energy, as described in Alonso et al. The main advantage of this type of models is that when they are fitted, they are really accurate, although they are too sensitive to the meteorological prediction error, which is usually amplified by these models. Hybrid models, finally, are a combination of these two models and they seem to be a promising approach that can outperform each of them individually.

PHOTOVOLTAIC SYSTEMS

A photovoltaic system, also PV system or solar power system, is a power system designed to supply usable solar power by means of photovoltaics. It consists of an arrangement of several components, including solar panels to absorb and convert sunlight into electricity, a solar inverter to convert the output from direct to alternating current, as well as mounting, cabling, and other electrical accessories to set up a working system. It may also use a solar tracking system to improve the system's overall performance and include an integrated battery solution, as prices for storage devices are expected to decline. Strictly speaking, a solar array only encompasses the ensemble of solar panels, the visible part of the PV system, and does not include all the other hardware, often summarized as balance of system (BOS). As PV systems convert light directly into electricity, such as concentrated solar power or solar thermal, used for heating and cooling.

PV systems range from small, rooftop-mounted or building-integrated systems with capacities from a few to several tens of kilowatts, to large utility-scale power stations of hundreds of megawatts. Nowadays, most PV systems are grid-connected, while off-grid or stand-alone systems account for a small portion of the market.

Operating silently and without any moving parts or environmental emissions, PV systems have developed from being niche market applications into a mature technology used for mainstream electricity generation. A rooftop system recoups the invested energy for its manufacturing and installation within 0.7 to 2 years and produces about 95 percent of net clean renewable energy over a 30-year service lifetime.

Due to the exponential growth of photovoltaics, prices for PV systems have rapidly declined in since their introduction. However, they vary by market and the size of the system. In 2014, prices for residential 5-kilowatt systems in the United States were around $3.29 per watt, while in the highly penetrated German market, prices for rooftop systems of up to 100 kW declined to €1.24 per watt. Nowadays, solar PV modules account for less than half of the system's overall cost, leaving the rest to the remaining BOS-components and to soft costs, which include customer acquisition, permitting, inspection and interconnection, installation labor and financing costs.

Modern System

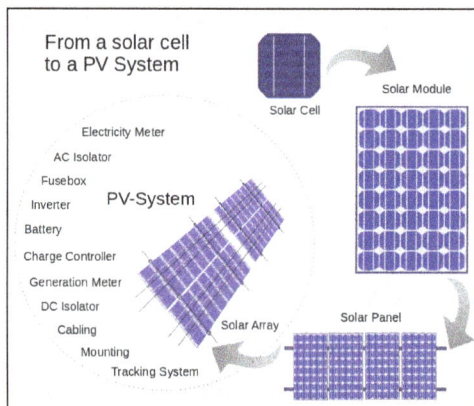

Diagram of the possible components of a photovoltaic system.

A photovoltaic system converts the sun's radiation, in the form of light, into usable electricity. It comprises the solar array and the balance of system components. PV systems can be categorized by various aspects, such as, grid-connected vs. stand alone systems, building-integrated vs. rack-mounted systems, residential vs. utility systems, distributed vs. centralized systems, rooftop vs. ground-mounted systems, tracking vs. fixed-tilt systems, and new constructed vs. retrofitted systems. Other distinctions may include, systems with microinverters vs. central inverter, systems using crystalline silicon vs. thin-film technology, and systems with modules from Chinese vs. European and U.S.-manufacturers.

About 99 percent of all European and 90 percent of all U.S. solar power systems are connected to the electrical grid, while off-grid systems are somewhat more common in Australia and South Korea. PV systems rarely use battery storage. This may change, as government incentives for distributed energy storage are implemented and investments in storage solutions gradually become economically viable for small systems. A typical residential solar array is rack-mounted on the roof, rather than integrated into the roof or facade of the building, which is significantly more expensive. Utility-scale solar power stations are ground-mounted, with fixed tilted solar panels rather than using expensive tracking devices. Crystalline silicon is the predominant material used in 90 percent of worldwide produced solar modules, while its rival thin-film has lost market-share. About 70 percent of all solar cells and modules are produced in China and Taiwan, only 5 percent by European and US-manufacturers. The installed capacity for both, small rooftop systems and large solar power stations is growing rapidly and in equal parts, although there is a notable trend towards utility-scale systems, as the focus on new installations is shifting away from Europe to sunnier regions, such as the Sunbelt in the U.S. which are less opposed to ground-mounted solar farms and cost-effectiveness is more emphasized by investors.

Driven by advances in technology and increases in manufacturing scale and sophistication, the cost of photovoltaics is declining continuously. There are several million PV systems distributed all over the world, mostly in Europe, with 1.4 million systems in Germany alone– as well as North America with 440,000 systems in the United States, The energy conversion efficiency of a conventional solar module increased from 15 to 20 percent since 2004 and a PV system recoups the energy needed for its manufacture in about 2 years. In exceptionally irradiated locations, or when thin-film technology is used, the so-called energy payback time decreases to one year or less. Net metering and financial incentives, such as preferential feed-in tariffs for solar-generated electricity, have also greatly supported installations of PV systems in many countries. The levelised cost of electricity from large-scale PV systems has become competitive with conventional electricity sources in an expanding list of geographic regions, and grid parity has been achieved in about 30 different countries.

As of 2015, the fast-growing global PV market is rapidly approaching the 200 GW mark – about 40 times the installed capacity in 2006. These systems currently contribute about 1 percent to worldwide electricity generation. Top installers of PV systems in terms of capacity are currently China, Japan and the United States, while half of the world's capacity is installed in Europe, with Germany and Italy supplying 7% to 8% of their respective domestic electricity consumption with solar PV. The International Energy Agency expects solar power to become the world's largest source of electricity by 2050, with solar photovoltaics and concentrated solar thermal contributing 16% and 11% to the global demand, respectively.

Grid-Connection

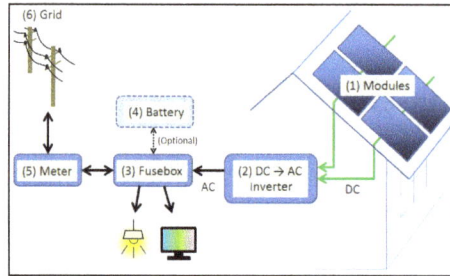

Schematics of a typical residential PV system.

A grid connected system is connected to a larger independent grid (typically the public electricity grid) and feeds energy directly into the grid. This energy may be shared by a residential or commercial building before or after the revenue measurement point, depending on whether the credited energy production is calculated independently of the customer's energy consumption (feed-in tariff) or only on the difference of energy (net metering). These systems vary in size from residential (2–10 kW$_p$) to solar power stations (up to 10s of MW$_p$). This is a form of decentralized electricity generation. Feeding electricity into the grid requires the transformation of DC into AC by a special, synchronising grid-tie inverter. In kilowatt-sized installations the DC side system voltage is as high as permitted (typically 1000 V except US residential 600 V) to limit ohmic losses. Most modules (60 or 72 crystalline silicon cells) generate 160 W to 300 W at 36 volts. It is sometimes necessary or desirable to connect the modules partially in parallel rather than all in series. An individual set of modules connected in series is known as a 'string'.

Scale of System

Photovoltaic systems are generally categorized into three distinct market segments: residential rooftop, commercial rooftop, and ground-mount utility-scale systems. Their capacities range from a few kilowatts to hundreds of megawatts. A typical residential system is around 10 kilowatts and mounted on a sloped roof, while commercial systems may reach a megawatt-scale and are generally installed on low-slope or even flat roofs. Although rooftop mounted systems are small and have a higher cost per watt than large utility-scale installations, they account for the largest share in the market. There is, however, a growing trend towards bigger utility-scale power plants, especially in the "sunbelt" region of the planet

Utility-Scale

Large utility-scale solar parks or farms are power stations and capable of providing an energy supply to large numbers of consumers. Generated electricity is fed into the transmission grid powered by central generation plants (grid-connected or grid-tied plant), or combined with one, or many, domestic electricity generators to feed into a small electrical grid (hybrid plant). In rare cases generated electricity is stored or used directly by island/standalone plant. PV systems are generally designed in order to ensure the highest energy yield for a given investment. Some large photovoltaic power stations such as Solar Star, Waldpolenz Solar Park and Topaz Solar Farm cover tens or hundreds of hectares and have power outputs up to hundreds of megawatts.

Perovo Solar Park in Ukraine.

Rooftop, Mobile and Portable

Rooftop system near Boston, USA.

A small PV system is capable of providing enough AC electricity to power a single home, or an isolated device in the form of AC or DC electric. Military and civilian Earth observation satellites, street lights, construction and traffic signs, electric cars, solar-powered tents, and electric aircraft may contain integrated photovoltaic systems to provide a primary or auxiliary power source in the form of AC or DC power, depending on the design and power demands. In 2013, rooftop systems accounted for 60 percent of worldwide installations. However, there is a trend away from rooftop and towards utility-scale PV systems, as the focus of new PV installations is also shifting from Europe to countries in the sunbelt region of the planet where opposition to ground-mounted solar farms is less accentuated. Portable and mobile PV systems provide electrical power independent of utility connections, for "off the grid" operation. Such systems are so commonly used on recreational vehicles and boats that there are retailers specializing in these applications and products specifically targeted to them. Since recreational vehicles (RV) normally carry batteries and operate lighting and other systems on nominally 12-volt DC power, RV systems normally operate in a voltage range that can charge 12-volt batteries directly, so addition of a PV system requires only panels, a charge controller, and wiring. Solar systems on recreation vehicles are usually constrained in wattage by the physical size of the RV's roof space.

Building-Integrated

BAPV wall near Barcelona, Spain.

In urban and suburban areas, photovoltaic arrays are often used on rooftops to supplement power use; often the building will have a connection to the power grid, in which
case the energy produced by the PV array can be sold back to the utility in some sort of
net metering agreement. Some utilities, use the rooftops of commercial customers and
telephone poles to support their use of PV panels. Solar trees are arrays that, as the name
implies, mimic the look of trees, provide shade, and at night can function as street lights.

Performance

Uncertainties in revenue over time relate mostly to the evaluation of the solar resource
and to the performance of the system itself. In the best of cases, uncertainties are typically 4% for year-to-year climate variability, 5% for solar resource estimation (in a horizontal plane), 3% for estimation of irradiation in the plane of the array, 3% for power
rating of modules, 2% for losses due to dirt and soiling, 1.5% for losses due to snow,
and 5% for other sources of error. Identifying and reacting to manageable losses is critical for revenue and O&M efficiency. Monitoring of array performance may be part of
contractual agreements between the array owner, the builder, and the utility purchas-
ing the energy produced. A method to create "synthetic days" using readily available
weather data and verification using the Open Solar Outdoors Test Field make it possible to predict photovoltaic systems performance with high degrees of accuracy. This
method can be used to then determine loss mechanisms on a local scale - such as those
from snow or the effects of surface coatings (e.g. hydrophobic or hydrophilic) on soiling
or snow losses. (Although in heavy snow environments with severe ground interference can result in annual losses from snow of 30%.) Access to the Internet has allowed
a further improvement in energy monitoring and communication. Dedicated systems
are available from a number of vendors. For solar PV systems that use microinverters
(panel-level DC to AC conversion), module power data is automatically provided. Some
systems allow setting performance alerts that trigger phone/email/text warnings when
limits are reached. These solutions provide data for the system owner and the installer.
Installers are able to remotely monitor multiple installations, and see at-a-glance the
status of their entire installed base.

Components

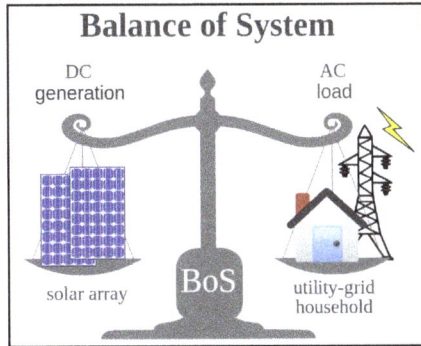

The balance of system components of a PV system (BOS) balance the
power-generating subsystem of the solar array (left side) with the power-using side
of the AC-household devices and the utility grid (right side).

A photovoltaic system for residential, commercial, or industrial energy supply consists of the solar array and a number of components often summarized as the balance of system (BOS). This term is synonymous with "Balance of plant" q.v. BOS-components include power-conditioning equipment and structures for mounting, typically one or more DC to AC power converters, also known as inverters, an energy storage device, a racking system that supports the solar array, electrical wiring and interconnections, and mounting for other components.

Optionally, a balance of system may include any or all of the following: renewable energy credit revenue-grade meter, maximum power point tracker (MPPT), battery system and charger, GPS solar tracker, energy management software, solar irradiance sensors, anemometer, or task-specific accessories designed to meet specialized requirements for a system owner. In addition, a CPV system requires optical lenses or mirrors and sometimes a cooling system.

The terms "solar array" and "PV system" are often incorrectly used interchangeably, despite the fact that the solar array does not encompass the entire system. Moreover, "solar panel" is often used as a synonym for "solar module", although a panel consists of a string of several modules. The term "solar system" is also an often used misnomer for a PV system.

Solar Array

Conventional c-Si solar cells, normally wired in series, are encapsulated in a solar module to protect them from the weather. The module consists of a tempered glass as cover, a soft and flexible encapsulant, a rear backsheet made of a weathering and fire-resistant material and an aluminium frame around the outer edge. Electrically connected and mounted on a supporting structure, solar modules build a string of modules, often called solar panel. A solar array consists of one or many such panels. A photovoltaic array, or solar array, is a linked collection of solar modules. The power that one module can produce is seldom enough to meet requirements of a home or a business, so the modules are linked together to form an array. Most PV arrays use an inverter to convert the DC power produced by the modules into alternating current that can power lights, motors, and other loads. The modules in a PV array are usually first connected in series to obtain the desired voltage; the individual strings are then connected in parallel to allow the system to produce more current. Solar panels are typically measured under STC (standard test conditions) or PTC (PVUSA test conditions), in watts. Typical panel ratings range from less than 100 watts to over 400 watts. The array rating consists of a summation of the panel ratings, in watts, kilowatts, or megawatts.

Module and Efficiency

A typical "150 watt" PV module is about a square meter in size. Such a module may be expected to produce 0.75 kilowatt-hour (kWh) every day, on average, after taking into account the weather and the latitude, for an insolation of 5 sun hours/day. In the last 10 years, the efficiency of average commercial wafer-based crystalline silicon modules increased from about 12% to 16% and CdTe module efficiency increased from 9% to 13% during same period. Module output and life degraded by increased temperature. Allowing ambient air to flow over, and if possible behind, PV modules reduces this problem. Effective module lives are typically 25 years or more. The payback period for an investment in a PV solar installation varies greatly and is typically less useful than a calculation of return on investment. While it is typically calculated to be between 10 and 20 years, the financial payback period can be far shorter with incentives.

Fixed tilt solar array in of crystalline silicon panels in Canterbury, New Hampshire, United States.

Solar array of a solar farm with a few thousand solar modules on the island of Majorca, Spain.

Due to the low voltage of an individual solar cell (typically ca. 0.5V), several cells are wired in series in the manufacture of a "laminate". The laminate is assembled into a protective weatherproof enclosure, thus making a photovoltaic module or solar panel. Modules may then be strung together into a photovoltaic array. In 2012, solar panels available for consumers have an efficiency of up to about 17%, while commercially available panels can go as far as 27%. It has been recorded that a group from The Fraunhofer Institute for Solar Energy Systems have created a cell that can reach 44.7% efficiency, which makes scientists' hopes of reaching the 50% efficiency threshold a lot more feasible.

Shading and Dirt

Photovoltaic cell electrical output is extremely sensitive to shading. When even a small portion of a cell, module, or array is shaded, with the remainder isin sunlight, the output falls dramatically due to internal 'short-circuiting' (the electrons reversing course through the shaded portion of the p-n junction). If the current drawn from the series string of cells is no greater than the current that can be produced by the shaded cell, the current (and so power) developed by the string is limited. If enough voltage is available from the other cells in a string, current will be forced through the cell by breaking down the junction in the shaded portion. This breakdown voltage in common cells is between 10 and 30 volts. Instead of adding to the power produced by the panel, the shaded cell absorbs power, turning it into heat. Since the reverse voltage of a shaded cell is much greater than the forward voltage of an illuminated cell, one shaded cell can absorb the power of many other cells in the string, disproportionately affecting panel output. For example, a shaded cell may drop 8 volts, instead of adding 0.5 volts, at a particular current level, thereby absorbing the power produced by 16 other cells. It is, thus important that a PV installation is not shaded by trees or other obstructions.

Several methods have been developed to determine shading losses from trees to PV systems over both large regions using LiDAR, but also at an individual system level using sketchup. Most modules have bypass diodes between each cell or string of cells that minimize the effects of shading and only lose the power of the shaded portion of the array. The main job of the bypass diode is to eliminate hot spots that form on cells that can cause further damage to the array, and cause fires. Sunlight can be absorbed by dust, snow, or other impurities at the surface of the module. This can reduce the light that strikes the cells. In general these losses aggregated over the year are small even for locations in Canada. Maintaining a clean module surface will increase output performance over the life of the module. Google found that cleaning flat mounted solar panels after 15 months increased their output by almost 100%, but that 5% tilted arrays were adequately cleaned by rainwater.

Insolation and Energy

Solar insolation is made up of direct, diffuse, and reflected radiation. The absorption factor of a PV cell is defined as the fraction of incident solar irradiance that is absorbed by the cell. At high noon on a cloudless day at the equator, the power of the sun is about 1 kW/m^2, on the Earth's surface, to a plane that is perpendicular to the sun's rays. As such, PV arrays can track the sun through each day to greatly enhance energy collection. However, tracking devices add cost, and require maintenance, so it is more common for PV arrays to have fixed mounts that tilt the array and face solar noon (approximately due south in the Northern Hemisphere or due north in the Southern Hemisphere). The tilt angle, from horizontal, can be varied for season, but if fixed, should be set to give optimal array output during the peak electrical demand portion of a typical year for a stand-alone system. This optimal module tilt angle is not necessarily identical to the tilt angle for maximum annual array energy output.

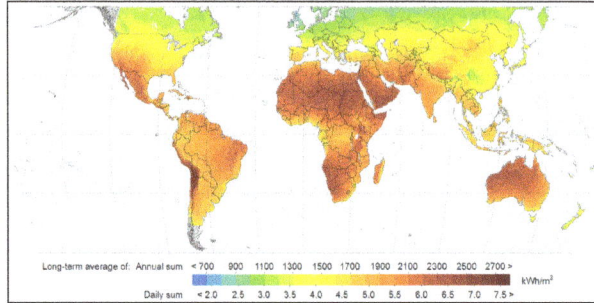

Global solar potential.

The optimization of the photovoltaic system for a specific environment can be complicated as issues of solar flux, soiling, and snow losses should be taken into effect. In addition, later work has shown that spectral effects can play a role in optimal photovoltaic material selection. For example, the spectral albedo can play a significant role in output depending on the surface around the photovoltaic system and the type of solar cell material. For the weather and latitudes of the United States and Europe, typical insolation ranges from 4 kWh/m²/day in northern climes to 6.5 kWh/m²/day in the sunniest regions. A photovoltaic installation in the southern latitudes of Europe or the United States may expect to produce 1 kWh/m²/day. A typical 1 kW photovoltaic installation in Australia or the southern latitudes of Europe or United States, may produce 3.5–5 kWh per day, dependent on location, orientation, tilt, insolation and other factors. In the Sahara desert, with less cloud cover and a better solar angle, one could ideally obtain closer to 8.3 kWh/m²/day provided the nearly ever present wind would not blow sand onto the units. The area of the Sahara desert is over 9 million km². 90,600 km², or about 1%, could generate as much electricity as all of the world's power plants combined.

Mounting

Modules are assembled into arrays on some kind of mounting system, which may be classified as ground mount, roof mount or pole mount. For solar parks a large rack is mounted on the ground, and the modules mounted on the rack. For buildings, many different racks have been devised for pitched roofs. For flat roofs, racks, bins and building integrated solutions are used. Solar panel racks mounted on top of poles can be stationary or moving, Side-of-pole mounts are suitable for situations where a pole has something else mounted at its top, such as a light fixture or an antenna. Pole mounting raises what would otherwise be a ground mounted array above weed shadows and livestock, and may satisfy electrical code requirements regarding inaccessibility of exposed wiring. Pole mounted panels are open to more cooling air on their underside, which

which increases performance. A multiplicity of pole top racks can be formed into a parking carport or other shade structure. A rack which does not follow the sun from left to right may allow seasonal adjustment up or down.

A 23-year-old, ground mounted PV system from the 1980s on a North Frisian Island, Germany. The modules conversion efficiency was only 12%.

Cabling

Due to their outdoor usage, solar cables are designed to be resistant against UV radiation and extremely high temperature fluctuations and are generally unaffected by the weather. Standards specifing the usage of electrical wiring in PV systems include the IEC 60364 by the International Electrotechnical Commission, "Solar photovoltaic (PV) power supply systems", the British Standard BS 7671, incorporating regulations relating to microgeneration and photovoltaic systems, and the US UL4703 standard, in subject 4703 "Photovoltaic Wire".

Tracker

A 1998 model of a passive solar tracker, viewed from underneath.

A solar tracking system tilts a solar panel throughout the day. Depending on the type of tracking system, the panel is either aimed directly at the sun or the brightest area of a partly clouded sky. Trackers greatly enhance early morning and late afternoon performance, increasing the total amount of power produced by a system by about 20–25% for a single axis tracker and about 30% or more for a dual axis tracker, depending on latitude. Trackers are effective in regions that receive a large portion of sunlight directly. In diffuse light (i.e. under cloud or fog), tracking has little or no value. Because most concentrated photovoltaics systems are very sensitive to the sunlight's angle, tracking systems allow them to produce useful power for more than a brief period each day. Tracking systems improve performance for two main reasons. First, when a solar panel is perpendicular to the sunlight, it receives more light on its surface than if it were angled. Second, direct light is used more efficiently than angled light. Special Anti-reflective coatings can improve solar panel efficiency for direct and angled light, somewhat reducing the benefit of tracking.

Trackers and sensors to optimise the performance are often seen as optional, but they can increase viable output by up to 45%. Arrays that approach or exceed one megawatt often use solar trackers. Considering clouds, and the fact that most of the world is not on the equator, and that the sun sets in the evening, the correct measure of solar power is insolation – the average number of kilowatt-hours per square meter per day. For the weather and latitudes of the United States and Europe, typical insolation ranges from 2.26 kWh/m²/day in northern climes to 5.61 kWh/m²/day in the sunniest regions.

For large systems, the energy gained by using tracking systems can outweigh the added complexity. For very large systems, the added maintenance of tracking is a substantial detriment. Tracking is not required for flat panel and low-concentration photovoltaic systems. For high-concentration photovoltaic systems, dual axis tracking is a necessity. Pricing trends affect the balance between adding more stationary solar panels versus having fewer panels that track.

As pricing, reliability and performance of single-axis trackers have improved, the systems have been installed in an increasing percentage of utility-scale projects. According to data from WoodMackenzie/GTM Research, global solar tracker shipments hit a record 14.5 gigawatts in 2017. This represents growth of 32 percent year-over-year, with similar or greater growth projected as large-scale solar deployment accelerates.

Inverter

Central inverter with AC and DC disconnects (on the side), monitoring gateway, transformer isolation and interactive LCD.

String inverter (left), generation meter, and AC disconnect (right). A modern 2013 installation in Vermont, United States.

Systems designed to deliver alternating current (AC), such as grid-connected applications need an inverter to convert the direct current (DC) from the solar modules to AC. Grid connected inverters must supply AC electricity in sinusoidal form, synchronized to the grid frequency, limit feed in voltage to no higher than the grid voltage and disconnect from the grid if the grid voltage is turned off. Islanding inverters need only produce regulated voltages and frequencies in a sinusoidal waveshape as no synchronisation or co-ordination with grid supplies is required.

A solar inverter may connect to a string of solar panels. In some installations a solar micro-inverter is connected at each solar panel. For safety reasons a circuit breaker is provided both on the AC and DC side to enable maintenance. AC output may be connected through an electricity meter into the public grid. The number of modules in the system determines the total DC watts capable of being generated by the solar array; however, the inverter ultimately governs the amount of AC watts that can be distributed for consumption. For example, a PV system comprising 11 kilowatts DC (kW_{DC}) worth of PV modules, paired with one 10-kilowatt AC (kW_{AC}) inverter, will be limited to the inverter's output of 10 kW. As of 2014, conversion efficiency for state-of-the-art converters reached more than 98 percent. While string inverters are used in residential to medium-sized commercial PV systems, central inverters cover the large commercial and utility-scale market. Market-share for central and string inverters are about 50 percent and 48 percent, respectively, with less than 2 percent for micro-inverters.

Maximum power point tracking (MPPT) is a technique that grid connected inverters use to get the maximum possible power from the photovoltaic array. In order to do so, the inverter's MPPT system digitally samples the solar array's ever changing power output and applies the proper resistance to find the optimal maximum power point.

Anti-islanding is a protection mechanism to immediately shut down the inverter, preventing it from generating AC power when the connection to the load no longer exists. This happens, for example, in the case of a blackout. Without this protection, the supply line would become an "island" with power surrounded by a "sea" of unpowered lines, as the solar array continues to deliver DC power during the power outage. Islanding is a hazard to utility workers, who may not realize that an AC circuit is still powered, and it may prevent automatic re-connection of devices.

Inverter/Converter Market in 2014				
Type	Power	Efficiency[a]	Market Share[b]	Remarks
String inverter	up to 100 kW_p [c]	98%	50%	Cost[b] €0.15 per watt-peak. Easy to replace.
Central inverter	above 100 kW_p	98.5%	48%	€0.10 per watt-peak. High reliability. Often sold along with a service contract.
Micro-inverter	module power range	90%–95%	1.5%	€0.40 per watt-peak. Ease of replacement concerns.

DC/DC con-verter Power opti-mizer	module power range	98.8%	n.a.	€0.40 per watt-peak. Ease of replacement concerns. Inverter is still needed. About 0.75 GW_p installed in 2013.

Battery

Although still expensive, PV systems increasingly use rechargeable batteries to store a surplus to be later used at night. Batteries used for grid-storage also stabilize the electrical grid by leveling out peak loads, and play an important role in a smart grid, as they can charge during periods of low demand and feed their stored energy into the grid when demand is high.

Common battery technologies used in today's PV systems include the valve regulated lead-acid battery– a modified version of the conventional lead–acid battery, nickel–cadmium and lithium-ion batteries. Compared to the other types, lead-acid batteries have a shorter lifetime and lower energy density. However, due to their high reliability, low self discharge as well as low investment and maintenance costs, they are currently the predominant technology used in small-scale, residential PV systems, as lithium-ion batteries are still being developed and about 3.5 times as expensive as lead-acid batteries. Furthermore, as storage devices for PV systems are stationary, the lower energy and power density and therefore higher weight of lead-acid batteries are not as critical as, for example, in electric transportation. Other rechargeable batteries considered for distributed PV systems include sodium–sulfur and vanadium redox batteries, two prominent types of a molten salt and a flow battery, respectively. In 2015, Tesla Motors launched the Powerwall, a rechargeable lithium-ion battery with the aim to revolutionize energy consumption.

PV systems with an integrated battery solution also need a charge controller, as the varying voltage and current from the solar array requires constant adjustment to prevent damage from overcharging. Basic charge controllers may simply turn the PV panels on and off, or may meter out pulses of energy as needed, a strategy called PWM or pulse-width modulation. More advanced charge controllers will incorporate MPPT logic into their battery charging algorithms. Charge controllers may also divert energy to some purpose other than battery charging. Rather than simply shut off the free PV energy when not needed, a user may choose to heat air or water once the battery is full.

Monitoring and Metering

The metering must be able to accumulate energy units in both directions, or two meters must be used. Many meters accumulate bidirectionally, some systems use two meters, but a unidirectional meter (with detent) will not accumulate energy from any resultant feed into the grid. In some countries, for installations over 30 kW_p a frequency and a voltage monitor with disconnection of all phases is required. This is done where more solar power is being generated than can be accommodated by the utility, and the excess can not either be exported or stored. Grid operators historically have needed to provide transmission lines and generation capacity. Now they need to also provide storage. This is normally hydro-storage, but other means of storage are used. Initially storage was used so that baseload generators could operate at full output. With variable renewable energy, storage is needed to allow power generation whenever it is available, and consumption whenever needed.

A Canadian electricity meter.

The two variables a grid operator have are storing electricity for *when* it is needed, or transmitting it to *where* it is needed. If both of those fail, installations over 30kWp can automatically shut down, although in practice all inverters maintain voltage regulation and stop supplying power if the load is inadequate. Grid operators have the option of curtailing excess generation from large systems, although this is more commonly done with wind power than solar power, and results in a substantial loss of revenue. Three-phase inverters have the unique option of supplying reactive power which can be advantageous in matching load requirements.

Photovoltaic systems need to be monitored to detect breakdown and optimize operation. There are several photovoltaic monitoring strategies depending on the output of the installation and its nature. Monitoring can be performed on site or remotely. It can measure production only, retrieve all the data from the inverter or retrieve all of the data from the communicating equipment (probes, meters, etc.). Monitoring tools can be dedicated to supervision only or offer additional functions. Individual inverters and battery charge controllers may include monitoring using manufacturer specific protocols and software. Energy metering of an inverter may be of limited accuracy and not suitable for revenue metering purposes. A third-party data acquisition system can monitor multiple inverters, using the inverter manufacturer's protocols, and also acquire weather-related information. Independent smart meters may measure the total energy production of a PV array system. Separate measures such as satellite image analysis or a solar radiation meter (a pyranometer) can be used to estimate total insolation for comparison. Data collected from a monitoring system can be displayed remotely over the World Wide Web, such as OSOTF.

Other Systems

Systems that are either highly specialized and uncommon or still an emerging new technology with limited significance. However, standalone or off-grid systems take a special place. They were the most common type of systems during the 1980s and 1990s, when PV technology was still very expensive and a pure niche market of small scale applications. Only in places where no electrical grid was available, they were economically viable. Although new stand-alone systems are still being deployed all around the world, their contribution to the overall installed photovoltaic capacity is decreasing. In Europe, off-grid systems account for 1 percent of installed capacity. In the United States, they account for about 10 percent. Off-grid systems are still common in Australia and South Korea, and in many developing countries.

CPV

Concentrator photovoltaic (CPV) in Catalonia, Spain.

Concentrator photovoltaics (CPV) and *high concentrator photovoltaic* (HCPV) systems use optical lenses or curved mirrors to concentrate sunlight onto small but highly efficient solar cells. Besides concentrating optics, CPV systems sometime use solar trackers and cooling systems and are more expensive.

Especially HCPV systems are best suited in location with high solar irradiance, concentrating sunlight up to 400 times or more, with efficiencies of 24–28 percent, exceeding those of regular systems. Various designs of systems are commercially available but not very common. However, ongoing research and development is taking place.

CPV is often confused with CSP (concentrated solar power) that does not use photovoltaics. Both technologies favor locations that receive much sunlight and are directly competing with each other.

Hybrid

A hybrid system combines PV with other forms of generation, usually a diesel generator. Biogas is also used. The other form of generation may be a type able to modulate power output as a function of demand. However more than one renewable form of energy may be used e.g. wind. The photovoltaic power generation serves to reduce the consumption of non renewable fuel. Hybrid systems are most often found on islands. Pellworm island in Germany and Kythnos island in Greece are notable examples (both are combined with wind). The Kythnos plant has reduced diesel consumption by 11.2%.

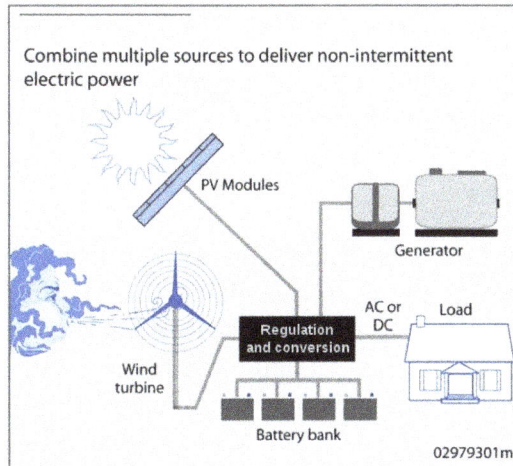

A wind-solar PV hybrid system.

In 2015, a case-study conducted in seven countries concluded that in all cases generating costs can be reduced by hybridising mini-grids and isolated grids. However, financing costs for such hybrids are crucial and largely depend on the ownership structure of the power plant. While cost reductions for state-owned utilities can be significant, the study also identified economic benefits to be insignificant or even negative for non-public utilities, such as independent power producers.

There has also been work showing that the PV penetration limit can be increased by deploying a distributed network of PV+CHP hybrid systems in the U.S. The temporal distribution of solar flux, electrical and heating requirements for representative U.S. single family residences were analyzed and the results clearly show that hybridizing CHP with PV can enable additional PV deployment above what is possible with a conventional centralized electric generation system. This theory was reconfirmed with numerical simulations using per second solar flux data to determine that the necessary battery backup to provide for such a hybrid system is possible with relatively small and inexpensive battery systems. In addition, large PV+CHP systems are possible for institutional buildings, which again provide back up for intermittent PV and reduce CHP runtime.

- PVT system (hybrid PV/T): Also known as photovoltaic thermal hybrid solar collectors convert solar radiation into thermal and electrical energy. Such a system combines a solar (PV) module with a solar thermal collector in a complementary way.

- CPVT system: A concentrated photovoltaic thermal hybrid (CPVT) system is similar to a PVT system. It uses concentrated photovoltaics (CPV) instead of conventional PV technology, and combines it with a solar thermal collector.

- CPV/CSP system: A novel solar CPV/CSP hybrid system has been proposed, combining concentrator photovoltaics with the non-PV technology of concentrated solar power (CSP), or also known as concentrated solar thermal.

- PV diesel system: It combines a photovoltaic system with a diesel generator. Combinations with other renewables are possible and include wind turbines.

Floating Solar Arrays

Floating solar arrays are PV systems that float on the surface of drinking water reservoirs, quarry lakes, irrigation canals or remediation and tailing ponds. These systems are called "floatovoltaics" when used only for electrical production or "aquavoltaics" when such systems are used to synergistically enhance aquaculture. A small number of such systems exist in France, India, Japan, South Korea, the United Kingdom, Singapore and the United States.

The systems are said to have advantages over photovoltaics on land. The cost of land is more expensive, and there are fewer rules and regulations for structures built on bodies of water not used for recreation. Unlike most land-based solar plants, floating arrays can be unobtrusive because they are hidden from public view. They achieve higher efficiencies than PV panels on land, because water cools the panels. The panels have a special coating to prevent rust or corrosion.

In May 2008, the Far Niente Winery in Oakville, California, pioneered the world's first floatovoltaic system by installing 994 solar PV modules with a total capacity of 477 kW onto 130 pontoons and floating them on the winery's irrigation pond. The primary benefit of such a system is that it avoids the need to sacrifice valuable land area that could be used for another purpose. In the case of the Far Niente Winery, it saved three-quarters of an acre that would have been required for a land-based system. Another benefit of a floatovoltaic system is that the panels are kept at a cooler temperature than they would be on land, leading to a higher efficiency of solar energy conversion. The floating PV array also reduces the amount of water lost through evaporation and inhibits the growth of algae.

Utility-scale floating PV farms are starting to be built. The multinational electronics and ceramics manufacturer Kyocera will develop the world's largest, a 13.4 MW farm on the reservoir above Yamakura Dam in Chiba Prefecture using 50,000 solar panels. Salt-water resistant floating farms are also being considered for ocean use, with experiments in Thailand. The largest so far announced floatovoltaic project is a 350 MW power station in the Amazon region of Brazil.

Direct Current Grid

DC grids are found in electric powered transport: railways trams and trolleybuses. A few pilot plants for such applications have been built, such as the tram depots in Hannover Leinhausen, using photovoltaic contributors and Geneva (Bachet de Pesay). The 150 kW$_p$ Geneva site feeds 600 V DC directly into the tram/trolleybus electricity network whereas before it provided about 15% of the electricity at its opening in 1999.

Standalone

An isolated mountain hut in Catalonia, Spain.

Solar parking meter in Edinburgh, Scotland.

A stand-alone or off-grid system is not connected to the electrical grid. Standalone systems vary widely in size and application from wristwatches or calculators to remote buildings or spacecraft. If the load is to be supplied independently of solar insolation, the generated power is stored and buffered with a battery. In non-portable applications where weight is not an issue, such as in buildings, lead acid batteries are most commonly used for their low cost and tolerance for abuse.

A charge controller may be incorporated in the system to avoid battery damage by excessive charging or discharging. It may also help to optimize production from the solar array using a maximum power point tracking technique (MPPT). However, in simple PV systems where the PV module voltage is matched to the battery voltage, the use of MPPT electronics is generally considered unnecessary, since the battery voltage is stable enough to provide near-maximum power collection from the PV module. In small devices (e.g. calculators, parking meters) only direct current (DC) is consumed. In larger systems (e.g. buildings, remote water pumps) AC is usually required. To convert the DC from the modules or batteries into AC, an inverter is used.

In agricultural settings, the array may be used to directly power DC pumps, without the need for an inverter. In remote settings such as mountainous areas, islands, or other places where a power grid is unavailable, solar arrays can be used as the sole source of electricity, usually by charging a storage battery. Stand-alone systems closely relate to microgeneration and distributed generation.

- Pico PV systems: The smallest, often portable photovoltaic systems are called pico solar PV systems, or pico solar. They mostly combine a rechargeable battery and charge controller, with a very small PV panel. The panel's nominal capacity is just a few watt-peak (1–10 W$_p$) and its area less than a tenth of a square meter, or one square foot, in size. A large range of different applications can be solar powered such as music players, fans, portable lamps, security lights, solar lighting kits, solar lanterns and street light, phone chargers, radios, or even small, seven-inch LCD televisions, that run on less than ten watts. As it is the case for power generation from pico hydro, pico PV systems are useful in small, rural communities that require only a small amount of electricity. Since the efficiency of many appliances have improved considerably, in particular due to the usage of LED lights and efficient rechargeable batteries, pico solar has become an affordable alternative, especially in the developing world. The metric prefix *pico-* stands for a *trillionth* to indicate the smallness of the system's electric power.

- Solar street lights: Solar street lights raised light sources which are powered by photovoltaic panels generally mounted on the lighting structure. The solar array of such off-grid PV system charges a rechargeable battery, which powers a fluorescent or LED lamp during the night. Solar street lights are stand-alone power systems, and have the advantage of savings on trenching, landscaping, and maintenance costs, as well as on the electric bills, despite their higher initial cost compared to conventional street lighting. They are designed with sufficiently large batteries to ensure operation for at least a week and even in the worst situation, they are expected to dim only slightly.

- Telecommunication and signaling: Solar PV power is ideally suited for telecommunication applications such as local telephone exchange, radio and TV broadcasting, microwave and other forms of electronic communication links. In most telecommunication application, storage batteries are already in use and the electrical system is basically DC. In hilly and mountainous terrain, radio and TV signals may not reach as they get blocked or reflected back due to undulating terrain. At these locations, low power transmitters are installed to receive and retransmit the signal for local population.

- Solar vehicles: Solar vehicle, whether ground, water, air or space vehicles may obtain some or all of the energy required for their operation from the sun. Surface vehicles generally require higher power levels than can be sustained by a practically sized solar array, so a battery assists in meeting peak power demand, and the solar array recharges it. Space vehicles have successfully used solar photovoltaic systems for years of operation, eliminating the weight of fuel or primary batteries.

- Solar pumps: One of the most cost effective solar applications is a solar powered pump, as it is far cheaper to purchase a solar panel than it is to run power lines. They often meet a need for water beyond the reach of power lines, taking the place of a windmill or windpump. One common application is the filling of livestock watering tanks, so that grazing cattle may drink. Another is the refilling of drinking water storage tanks on remote or self-sufficient homes.

- Spacecraft: Solar panels on spacecraft have been one of the first applications of photovoltaics since the launch of Vanguard 1 in 1958, the first satellite to use solar cells. Contrary to Sputnik, the first artificial satellite to orbit the planet, that ran out of batteries within 21 days due to the lack of solar-power, most modern communications satellites and space probes in the inner solar system rely on the use of solar panels to derive electricity from sunlight.

- Do it yourself community: With agrowing interest in environmentally friendly green energy, hobbyists in the DIY-community have endeavored to build their own solar PV systems from kits or partly DIY. Usually, the DIY-community uses inexpensive or high efficiency systems (such as those with solar tracking) to generate their own power. As a result, the DIY-systems often end up cheaper than their commercial counterparts. Often the system is also connected to the regular power grid, using net metering instead of a battery for backup. These systems usually generate power amount of ~2 kW or less. Through the internet, the community is now able to obtain plans to (partly) construct the system and there is a growing trend toward building them for domestic requirements.

A mobile charging station for electric vehicles in France.

Solar pathway lighting in winter (Steamboat Springs, US).

A solar sewage treatment plant in Santuari de Lluc, Spain.

System Cost 2013

In its 2014 edition of the "Technology Roadmap: Solar Photovoltaic Energy" report, the International Energy Agency (IEA) published prices in US$ per watt for residential, commercial and utility-scale PV systems for eight major markets in 2013.

Table: Typical PV system prices in 2013 in selected countries (USD).

USD/W	Australia	China	France	Germany	Italy	Japan	United Kingdom	United States
Residential	1.8	1.5	4.1	2.4	2.8	4.2	2.8	4.9
Commercial	1.7	1.4	2.7	1.8	1.9	3.6	2.4	4.5
Utility-scale	2.0	1.4	2.2	1.4	1.5	2.9	1.9	3.3

Learning Curve

Photovoltaic systems demonstrate a learning curve in terms of Levelized Cost of Electricity (LCOE), reducing its cost per kWh by 32.6% for every doubling of capacity. From the data of LCOE and cumulative installed capacity from International Renewable Energy Agency (IRENA) from 2010 to 2017, the learning curve equation for photovoltaic systems is given as

$$LCOE_{photovoltaic} = 151.46 Capacity^{-0.57}$$

- LCOE : levelized cost of electricity (in USD/kWh)

- Capacity : cumulative installed capacity of photovoltaic systems (in MW)

Limitations

Pollution and Energy in PV Production

PV has been a well-known method of generating clean, emission free electricity. PV systems are often made of PV modules and inverter (changing DC to AC). PV modules are mainly made of PV cells, which has no fundamental difference to the material for making computer chips. The process of producing PV cells (computer chips) is energy intensive and involves highly poisonous and environmental toxic chemicals. There are few PV manufacturing plants around the world producing PV modules with energy produced from PV. This measure greatly reduces the carbon footprint during the manufacturing process. Managing the chemicals used in the manufacturing process is subject to the factories' local laws and regulations.

Impact on Electricity Network

With the increasing levels of rooftop photovoltaic systems, the energy flow becomes 2-way. When there is more local generation than consumption, electricity is exported to the grid. However, electricity network traditionally is not designed to deal with the 2- way energy transfer. Therefore, some technical issues may occur. For example, in Queensland Australia, there have been more than 30% of households with rooftop PV by the end of 2017. The famous Californian 2020 duck curve appears very often for a lot of communities from 2015 onwards. An over-voltage issue may come out as the electricity flows back to the network. There are solutions to manage the over voltage issue, such as regulating PV inverter power factor, new voltage and energy control equipment at electricity distributor level, re-conductor the electricity wires, demand side management, etc. There are often limitations and costs related to these solutions.

Implication onto Electricity Bill Management and Energy Investment

Customers have different specific situations, e.g. different comfort/convenience needs, different electricity tariffs, or different usage patterns. An electricity tariff may have a few elements, such as daily access and metering charge, energy charge (based on kWh, MWh) or peak demand charge (e.g. a price for the highest 30min energy consumption in a month). PV is a promising option for reducing energy charge when electricity price is reasonably high and continuously increasing, such as in Australia and Germany. However, for sites with peak demand charge in place, PV may be less attractive if peak demands mostly occur in the late afternoon to early evening, for example residential communities. Overall, energy investment is largely an economic decision and investment decisions are based on systematical evaluation of options in operational improvement, energy efficiency, onsite generation and energy storage.

SOLAR PANELS

Photovoltaic solar panels absorb sunlight as a source of energy to generate direct current electricity. A photovoltaic (PV) module is a packaged, connected assembly of photovoltaic solar cells available in different voltages and wattages. Photovoltaic modules constitute the photovoltaic array of a photovoltaic system that generates and supplies solar electricity in commercial and residential applications.

Solar PV modules (top) and two solar hot water panels (bottom) mounted on rooftops.

The most common application of solar energy collection outside agriculture is solar water heating systems.

Theory and Construction

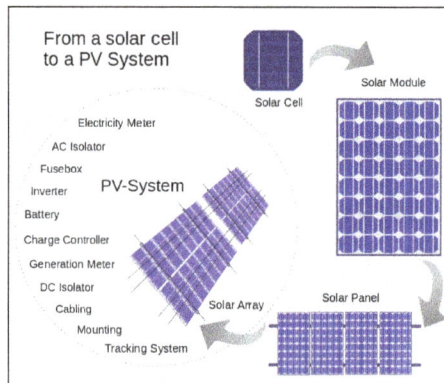

From a solar cell to a PV system.

Photovoltaic modules use light energy (photons) from the Sun to generate electricity through the photovoltaic effect. The majority of modules use wafer-based crystalline silicon cells or thin-film cells. The structural (load carrying) member of a module can either be the top layer or the back layer. Cells must also be protected from mechanical damage and moisture. Most modules are rigid, but semi-flexible ones based on thin-film cells are also available. The cells are connected electrically in series, one to another to a desired voltage, and then in parallel to increase amperage. The voltage and amperage of the module are multiplied to create the wattage of the module.

A PV junction box is attached to the back of the solar panel and it is its output interface. Externally, most of photovoltaic modules use MC4 connectors type to facilitate easy weatherproof connections to the rest of the system. Also, USB power interface can be used.

Module electrical connections are made in series to achieve a desired output voltage or in parallel to provide a desired current capability (amperes) of the solar panel or the PV system. The conducting wires that take the current off the modules are sized according to the ampacity and may contain silver, copper or other non-magnetic conductive transition metals. Bypass diodes may be incorporated or used externally, in case of partial module shading, to maximize the output of module sections still illuminated.

Some special solar PV modules include concentrators in which light is focused by lenses or mirrors onto smaller cells. This enables the use of cells with a high cost per unit area (such as gallium arsenide) in a cost-effective way. Solar panels also use metal frames consisting of racking components, brackets, reflector shapes, and troughs to better support the panel structure.

Efficiencies

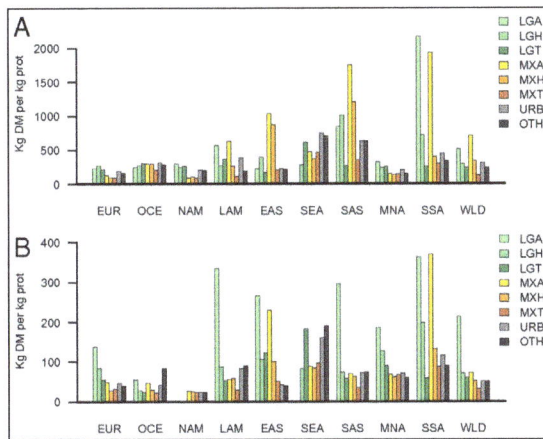

Reported timeline of champion solar module energy conversion efficiencies since 1988.

Each module is rated by its DC output power under standard test conditions (STC), and typically ranges from 100 to 365 Watts (W). The efficiency of a module determines the area of a module given the same rated output – an 8% efficient 230 W module will have twice the area of a 16% efficient 230 W module. There are a few commercially available solar modules that exceed efficiency of 24%.

Depending on construction, photovoltaic modules can produce electricity from a range of frequencies of light, but usually cannot cover the entire solar range (specifically, ultraviolet, infrared and low or diffused light). Hence, much of the incident sunlight energy is wasted by solar modules, and they can give far higher efficiencies if illuminated with monochromatic light. Therefore, another design concept is to split the light into six to eight different wavelength ranges that will produce a different color of light, and direct the beams onto different cells tuned to those ranges. This has been projected to be capable of raising efficiency by 50%.

A single solar module can produce only a limited amount of power; most installations contain multiple modules adding voltages or current to the wiring and PV system. A photovoltaic system typically includes an array of photovoltaic modules, an inverter, a battery pack for energy storage, charge controller, interconnection wiring, circuit breakers, fuses, disconnect switches, voltage meters, and optionally a solar tracking mechanism. Equipment is carefully selected to optimize output, energy storage, reduce power loss during power transmission, and conversion from direct current to alternating current.

Scientists from Spectrolab, a subsidiary of Boeing, have reported development of multi-junction solar cells with an efficiency of more than 40%, a new world record for solar photovoltaic cells. The Spectrolab scientists also predict that concentrator solar cells could achieve efficiencies of more than 45% or even 50% in the future, with theoretical efficiencies being about 58% in cells with more than three junctions.

Currently, the best achieved sunlight conversion rate (solar module efficiency) is around 21.5% in new commercial products typically lower than the efficiencies of their cells in isolation. The most efficient mass-produced solar moduleshave power density values of up to 175 W/m² (16.22 W/ft²).

Research by Imperial College, London has shown that the efficiency of a solar panel can be improved by studding the light-receiving semiconductor surface with aluminum nanocylinders similar to the ridges on Lego blocks. The scattered light then travels along a longer path in the semiconductor which means that more photons can be absorbed and converted into current. Although these nanocylinders have been used previously (aluminum was preceded by gold and silver), the light scattering occurred in the near infrared region and visible light was absorbed strongly. Aluminum was found to have absorbed the ultraviolet part of the spectrum, while the visible and near infrared parts of the spectrum were found to be scattered by the aluminum surface. This, the research argued, could bring down the cost significantly and improve the efficiency as aluminum is more abundant and less costly than gold and silver. The research also noted that the increase in current makes thinner film solar panels technically feasible without "compromising power conversion efficiencies, thus reducing material consumption".

- Efficiencies of solar panel can be calculated by MPP (maximum power point) value of solar panels.

- Solar inverters convert the DC power to AC power by performing MPPT process: solar inverter samples the output Power (I-V curve) from the solar cell and applies the proper resistance (load) to solar cells to obtain maximum power.

- MPP (Maximum power point) of the solar panel consists of MPP voltage (V mpp) and MPP current (I mpp): it is a capacity of the solar panel and the higher value can make higher MPP.

Micro-inverted solar panels are wired in parallel, which produces more output than normal panels which are wired in series with the output of the series determined by the lowest performing panel (this is known as the "Christmas light effect"). Micro-inverters work independently so each panel contributes its maximum possible output given the available sunlight.

Technology

Global Market Share by PV Technology
from 1990 to 2013

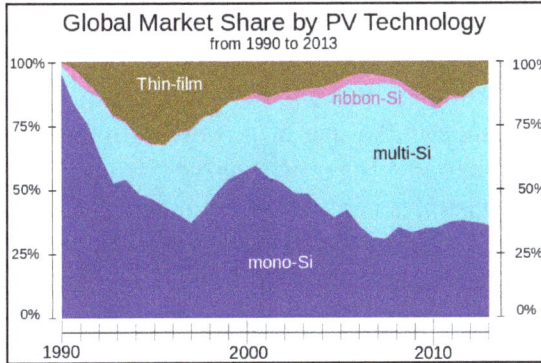

Market-share of PV technologies since 1990.

Most solar modules are currently produced from crystalline silicon (c-Si) solar cells made of multicrystalline and monocrystalline silicon. In 2013, crystalline silicon accounted for more than 90 percent of worldwide PV production, while the rest of the overall market is made up of thin-film technologies using cadmium telluride, CIGS and amorphous silicon.

Emerging, third generation solar technologies use advanced thin-film cells. They produce a relatively high-efficiency conversion for the low cost compared to other solar technologies. Also, high-cost, high-efficiency, and close-packed rectangular multi-junction (MJ) cells are preferably used in solar panels on spacecraft, as they offer the highest ratio of generated power per kilogram lifted into space. MJ-cells are compound semiconductors and made of gallium arsenide (GaAs) and other semiconductor materials. Another emerging PV technology using MJ-cells is concentrator photovoltaics (CPV).

Thin Film

In rigid thin-film modules, the cell and the module are manufactured in the same production line. The cell is created on a glass substrate or superstrate, and the electrical connections are created *in situ*, a so-called "monolithic integration". The substrate or superstrate is laminated with an encapsulant to a front or back sheet, usually another sheet of glass. The main cell technologies in this category are CdTe, or a-Si, or a-Si+uc-Si tandem, or CIGS (or variant). Amorphous silicon has a sunlight conversion rate of 6–12%

Flexible thin film cells and modules are created on the same production line by depositing the photoactive layer and other necessary layers on a flexible substrate. If the substrate is an insulator (e.g. polyester or polyimide film) then monolithic integration can be used. If it is a conductor then another technique for electrical connection must be used. The cells are assembled into modules by laminating them to a transparent colourless fluoropolymer on the front side (typically ETFE or FEP) and a polymer suitable for bonding to the final substrate on the other side.

Smart Solar Modules

Several companies have begun embedding electronics into PV modules. This enables performing maximum power point tracking (MPPT) for each module individually, and the measurement of performance data for monitoring and fault detection at module level. Some of these solutions make use of power optimizers, a DC-to-DC converter technology developed to maximize the power harvest from solar photovoltaic systems. As of about 2010, such electronics can also compensate for shading effects, wherein a shadow falling across a section of a module causes the electrical output of one or more strings of cells in the module to fall to zero, but not having the output of the entire module fall to zero.

Performance and Degradation

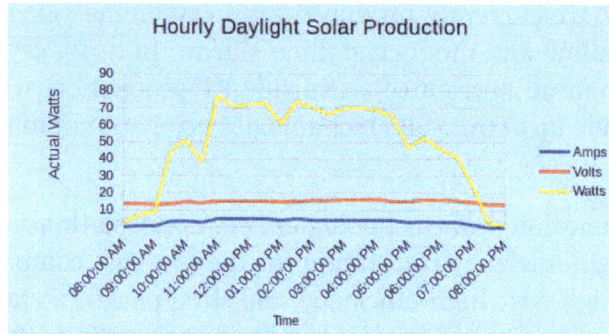

Module performance is generally rated under standard test conditions (STC): irradiance of 1,000 W/m², solar spectrum of AM 1.5 and module temperature at 25°C. The actual voltage and current output of the module changes as lighting, temperature and load conditions change, so there is never one specific voltage, current, or wattage at which the module operates. Performance varies depending on time of day, amount of solar insolation, direction and tilt of modules, cloud cover, shading, temperature, geographic location, and day of the year.

For optimum performance a solar panel needs to be made of similar modules oriented in the same direction perpendicular towards direct sunlight. The path of the sun varies by latitude and day of the year and can be studied using a sundial or a sunchart and tracked using a solar tracker. Differences in voltage or current of modules may affect the overall performance of a panel. Bypass diodes are used to circumvent broken or shaded panels to optimize output.

Electrical characteristics include nominal power (P_{MAX}, measured in W), open circuit voltage (V_{OC}), short circuit current (I_{SC}, measured in amperes), maximum power voltage (V_{MPP}), maximum power current (I_{MPP}), peak power, (watt-peak, W_p), and module efficiency (%).

Nominal voltage refers to the voltage of the battery that the module is best suited to charge; this is a leftover term from the days when solar modules were only used to charge batteries. Nominal voltage allows users, at a glance, to make sure the module is compatible with a given system.

Open circuit voltage or V_{OC} is the maximum voltage that the module can produce when not connected to an electrical circuit or system. V_{OC} can be measured with a voltmeter directly on an illuminated module's terminals or on its disconnected cable.

The peak power rating, W_p, is the maximum output under standard test conditions (not the maximum possible output). Typical modules, which could measure approximately 1 m × 2 m or 3 ft 3 in × 6 ft 7 in, will be rated from as low as 75 W to as high as 350 W, depending on their efficiency. At the time of testing, the test modules are binned according to their test results, and a typical manufacturer might rate their modules in 5 W increments, and either rate them at +/- 3%, +/-5%, +3/-0% or +5/-0%.

Solar water heater.

The ability of solar modules to withstand damage by rain, hail, heavy snow load, and cycles of heat and cold varies by manufacturer, although most solar panels on the U.S. market are UL listed, meaning they have gone through testing to withstand hail. Many crystalline silicon module manufacturers offer a limited warranty that guarantees electrical production for 10 years at 90% of rated power output and 25 years at 80%.

Potential induced degradation (also called PID) is a potential induced performance degradation in crystalline photovoltaic modules, caused by so-called stray currents. This effect may cause power loss of up to 30%.

The largest challenge for photovoltaic technology is said to be the purchase price per watt of electricity produced. New materials and manufacturing techniques continue to improve the price to power performance. The problem resides in the enormous activation energy that must be overcome for a photon to excite an electron for harvesting purposes. Advancements in photovoltaic technologies have brought about the process of "doping" the silicon substrate to lower the activation energy thereby making the panel more efficient in converting photons to retrievable electrons.

Chemicals such as boron (p-type) are applied into the semiconductor crystal in order to create donor and acceptor energy levels substantially closer to the valence and conductor bands. In doing so, the addition of boron impurity allows the activation energy to decrease 20 fold from 1.12 eV to 0.05 eV. Since the potential difference (E_B) is so low, the boron is able to thermally ionize at room temperatures. This allows for free energy carriers in the conduction and valence bands thereby allowing greater conversion of photons to electrons.

Maintenance

Solar panel conversion efficiency, typically in the 20% range, is reduced by dust, grime, pollen, and other particulates that accumulate on the solar panel. "A dirty solar panel can reduce its power capabilities by up to 30% in high dust/pollen or desert areas", says Seamus Curran, associate professor of physics at the University of Houston and director of the Institute for NanoEnergy, which specializes in the design, engineering, and assembly of nanostructures.

Paying to have solar panels cleaned is often not a good investment; researchers found panels that had not been cleaned, or rained on, for 145 days during a summer drought in California, lost only 7.4% of their efficiency. Overall, for a typical residential solar system of 5 kW, washing panels halfway through the summer would translate into a mere $20 gain in electricity production until the summer drought ends—in about 2 ½ months. For larger commercial rooftop systems, the financial losses are bigger but still rarely enough to warrant the cost of washing the panels. On average, panels lost a little less than 0.05% of their overall efficiency per day.

Recycling

Most parts of a solar module can be recycled including up to 95% of certain semiconductor materials or the glass as well as large amounts of ferrous and non-ferrous metals. Some private companies and non-profit organizations are currently engaged in take-back and recycling operations for end-of-life modules.

Recycling possibilities depend on the kind of technology used in the modules:

- Silicon based modules: Aluminum frames and junction boxes are dismantled manually at the beginning of the process. The module is then crushed in a mill and the different fractions are separated - glass, plastics and metals. It is possible to recover more than 80% of the incoming weight. This process can be performed by flat glass recyclers since morphology and composition of a PV module is similar to those flat glasses used in the building and automotive industry. The recovered glass for example is readily accepted by the glass foam and glass insulation industry.

- Non-silicon based modules: They require specific recycling technologies such as the use of chemical baths in order to separate the different semiconductor materials. For cadmium telluride modules, the recycling process begins by crushing the module and subsequently separating the different fractions.

This recycling process is designed to recover up to 90% of the glass and 95% of the semiconductor materials contained. Some commercial-scale recycling facilities have been created in recent years by private companies. For aluminium flat plate reflector: the trendiness of the reflectors has been brought up by fabricating them using a thin layer (around 0.016 mm to 0.024 mm) of Aluminum coating present inside the non-recycled plastic food packages.

Since 2010, there is an annual European conference bringing together manufacturers, recyclers and researchers to look at the future of PV module recycling.

Connectors

Outdoor solar panels usually includes MC4 connectors. Automotive solar panels also can include car lighter and USB adapter. Indoor panels (including solar pv glasses, thin films and windows) can integrate microinverter (AC Solar panels).

Applications

There are many practical applications for the use of solar panels or photovoltaics. It can first be used in agriculture as a power source for irrigation. In health care solar panels can be used to refrigerate medical supplies. It can also be used for infrastructure. PV modules are used in photovoltaic systems and include a large variety of electric devices:

- Photovoltaic power stations.
- Rooftop solar PV systems.
- Standalone PV systems.
- Solar hybrid power systems.
- Concentrated photovoltaics.
- Solar planes.
- Solar-pumped lasers.
- Solar vehicles.
- Solar panels on spacecrafts and space stations.

Limitations

Pollution and Energy in Production

Solar panel has been a well-known method of generating clean, emission free electricity. However, it produces only direct current electricity (DC), which is not what normal appliances use. Solar photovoltaic systems (solar PV systems) are often made of solar PV panels (modules) and inverter (changing DC to AC). Solar PV panels are mainly made of solar photovoltaic cells, which has no fundamental difference to the material for making computer chips. The process of producing solar PV cells (computer chips) is energy intensive and involves highly poisonous and environmental toxic chemicals. There are few solar PV manufacturing plants around the world producing PV modules with energy produced from PV. This measure greatly reduces the carbon footprint during the manufacturing process. Managing the chemicals used in the manufacturing process is subject to the factories' local laws and regulations.

Impact on Electricity Network

With the increasing levels of rooftop photovoltaic systems, the energy flow becomes 2-way. When there is more local generation than consumption, electricity is exported to the grid. However, electricity network traditionally is not designed to deal with the 2- way energy transfer. Therefore, some technical issues may occur. For example in Queensland Australia, there have been more than 30% of households with rooftop PV by the end of 2017. The famous Californian 2020 duck curve appears very often for a lot of communities from 2015 onwards. An over-voltage issue may come out as the electricity flows from these PV households back to the network. There are solutions to manage the over voltage issue, such as regulating PV inverter power factor, new voltage and energy control equipment at electricity distributor level, re-conducting the electricity wires, demand side management, etc. There are often limitations and costs related to these solutions.

Implication onto Electricity Bill Management and Energy Investment

There is no silver bullet in electricity or energy demand and bill management, because customers (sites) have different specific situations, e.g. different comfort/convenience needs, different electricity tariffs, or different usage patterns. Electricity tariff may have a few elements, such as daily access and metering charge, energy charge (based on kWh, MWh) or peak demand charge (e.g. a price for the highest 30min energy consumption in a month). PV is a promising option for reducing energy charge when electricity price is reasonably high and continuously increasing, such as in Australia and Germany. However for sites with peak demand charge in place, PV may be less attractive if peak demands mostly occur in the late afternoon to early evening, for example residential communities. Overall, energy investment is largely an economical decision and it is better to make investment decisions based on systematical evaluation of options in operational improvement, energy efficiency, onsite generation and energy storage.

AIR MASS

The air mass coefficient defines the direct optical path length through the Earth's atmosphere, expressed as a ratio relative to the path length vertically upwards, i.e.

at the zenith. The air mass coefficient can be used to help characterize the solar spectrum after solar radiation has traveled through the atmosphere. The air mass coefficient is commonly used to characterize the performance of solar cells under standardized conditions, and is often referred to using the syntax "AM" followed by a number. "AM1.5" is almost universal when characterizing terrestrial power-generating panels.

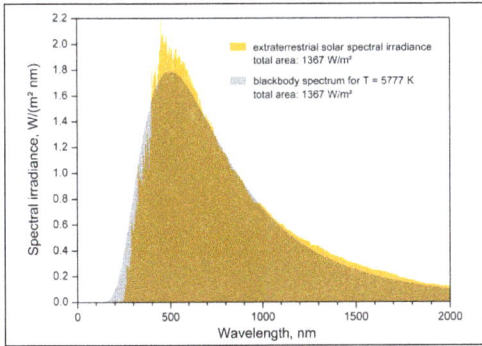

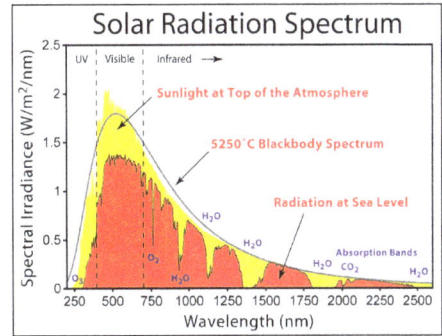

The effective temperature, or black body temperature, of the Sun (5777 K) is the temperature a black body of the same size must have to yield the same total emissive power.

Solar irradiance spectrum above atmosphere and at surface.

Solar radiation closely matches a black body radiator at about 5,800 K. As it passes through the atmosphere, sunlight is attenuated by scattering and absorption; the more atmosphere through which it passes, the greater the attenuation.

As the sunlight travels through the atmosphere, chemicals interact with the sunlight and absorb certain wavelengths changing the amount of short-wavelength light reaching the Earth's surface. A more active component of this process is water vapor, which results in a wide variety of absorption bands at many wavelengths, while molecular nitrogen, oxygen and carbon dioxide add to this process. By the time it reaches the Earth's surface, the spectrum is strongly confined between the far infrared and near ultraviolet.

Atmospheric scattering plays a role in removing higher frequencies from direct sunlight and scattering it about the sky. This is why the sky appears blue and the sun yellow — more of the higher-frequency blue light arrives at the observer via indirect scattered paths; and less blue light follows the direct path, giving the sun a yellow tinge. The greater the distance in the atmosphere through which the sunlight travels, the greater this effect, which is why the sun looks orange or red at dawn and sundown when the sunlight is travelling very obliquely through the atmosphere — progressively more of the blues and greens are removed from the direct rays, giving an orange or red appearance to the sun; and the sky appears pink — because the blues and greens are scattered over such long paths that they are highly attenuated before arriving at the observer, resulting in characteristic pink skies at dawn and sunset.

For a path length L through the atmosphere, and solar radiation incident at angle z relative to the normal to the Earth's surface, the air mass coefficient is:

$$AM = \frac{L}{L_o}$$

where L_o is the path length at zenith (i.e. normal to the Earth's surface) at sea level.

The air mass number is thus dependent on the Sun's elevation path through the sky and therefore varies with time of day and with the passing seasons of the year, and with the latitude of the observer.

Calculation

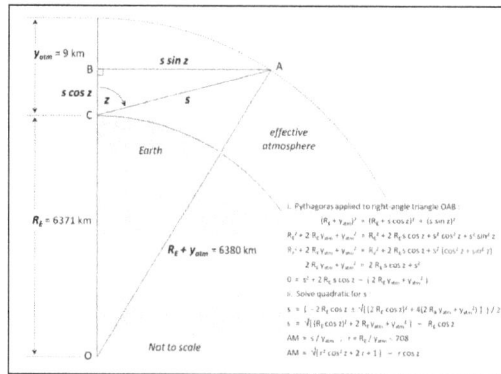

Atmospheric effects on optical transmission can be modelled as if the atmosphere is concentrated in approximately the lower 9 km.

A first-order approximation for air mass is given by,

$$AM \approx \frac{1}{\cos z}$$

where z is the zenith angle in degrees.

The above approximation overlooks the atmosphere's finite height, and predicts an infinite air mass at the horizon. However, it is reasonably accurate for values of z up to around 75°. A number of refinements have been proposed to more accurately model the path thickness towards the horizon, such as that proposed by Kasten and Young (1989):

$$AM = \frac{1}{\cos\ z + 0.50572(96.07995 - z)^{-1.6364}}$$

At sea level the air mass towards the horizon (= 90°) is approximately 38.

Modelling the atmosphere as a simple spherical shell provides a reasonable approximation:

$$AM = \sqrt{(r\cos z)^2 + 2r + 1} - r\cos z$$

where the radius of the Earth R_E = 6371 km, the effective height of the atmosphere y_{atm} ≈ 9 km, and their ratio $r = R_E / y_{atm}$ ≈ 708.

These models are compared in the table below:

Estimates of airmass coefficient at sea level			
z	Flat Earth	Kasten & Young	Spherical shell
degree	(A.1)	(A.2)	(A.3)
0°	1.0	1.0	1.0
60°	2.0	2.0	2.0
70°	2.9	2.9	2.9
75°	3.9	3.8	3.8
80°	5.8	5.6	5.6
85°	11.5	10.3	10.6
88°	28.7	19.4	20.3
90°		37.9	37.6

This implies that for these purposes the atmosphere can be considered to be effectively concentrated into around the bottom 9 km, i.e. essentially all the atmospheric effects are due to the atmospheric mass in the lower half of the Troposphere. This is a useful and simple model when considering the atmospheric effects on solar intensity.

Cases

- AM0: The spectrum outside the atmosphere, approximated by the 5,800 K black body, is referred to as "AM0", meaning "zero atmospheres". Solar cells used for space power applications, like those on communications satellites are generally characterized using AM0.

- AM1: The spectrum after travelling through the atmosphere to sea level with the sun directly overhead is referred to, by definition, as "AM1". This means "one atmosphere". AM1 ($z = 0°$) to AM1.1 ($z = 25°$) is a useful range for estimating performance of solar cells in equatorial and tropical regions.

- AM1.5: Solar panels do not generally operate under exactly one atmosphere's thickness: if the sun is at an angle to the Earth's surface the effective thickness will be greater. Many of the world's major population centres, and hence solar installations and industry, across Europe, China, Japan, the United States of America and elsewhere (including northern India, southern Africa and Australia) lie in temperate latitudes. An AM number representing the spectrum at mid-latitudes is therefore much more common.

- "AM1.5", 1.5 atmosphere thickness, corresponds to a solar zenith angle of z =48.2°. While the summertime AM number for mid-latitudes during the middle parts of the day is less than 1.5, higher figures apply in the morning and evening and at other times of the year. Therefore, AM1.5 is useful to represent the overall yearly average for mid-latitudes. The specific value of 1.5 has been selected in the 1970s for standardization purposes, based on an analysis of solar irradiance data in the conterminous United States. Since then, the solar industry has been using AM1.5 for all standardized testing or rating of terrestrial solar cells or modules, including those used in concentrating systems. The latest AM1.5 standards pertaining to photovoltaic applications are the ASTM G-173 and IEC 60904, all derived from simulations obtained with the SMARTS code

- AM2~3: AM2 (=60°) to AM3 (z=70°) is a useful range for estimating the overall average performance of solar cells installed at high latitudes such as in northern Europe. Similarly AM2 to AM3 is useful to estimate wintertime performance in temperate latitudes, e.g. airmass coefficient is greater than 2 at all hours of the day in winter at latitudes as low as 37°.

- AM38: AM38 is generally regarded as being the airmass in the horizontal direction (z=90°) at sea level. However, in practice there is a high degree of variability in the solar intensity received at angles close to the horizon.

- The *relative* air mass is only a function of the sun's zenith angle, and therefore does not change with local elevation. Conversely, the *absolute* air mass, equal to the relative air mass multiplied by the local atmospheric pressure and divided by the standard (sea-level) pressure, decreases with elevation above sea level. For solar panels installed at high altitudes, e.g. in an Altiplano region, it is possible to use a lower absolute AM numbers than for the corresponding latitude at sea level: AM numbers less than 1 towards the equator, and correspondingly lower numbers than listed above for other latitudes. However, this approach is approximate and not recommended. It is best to simulate the actual spectrum based on the relative air mass (e.g. 1.5) and the *actual* atmospheric conditions for the specific elevation of the site under scrutiny.

Solar Intensity

Solar intensity at the collector reduces with increasing airmass coefficient, but due to the complex and variable atmospheric factors involved, not in a simple or linear fashion. For example, almost all high energy radiation is removed in the upper atmosphere (between AM0 and AM1) and so AM2 is not twice as bad as AM1.

Furthermore, there is great variability in many of the factors contributing to atmospheric attenuation, such as water vapor, aerosols, photochemical smog and the effects of temperature inversions. Depending on level of pollution in the air, overall attenuation can change by up to ±70% towards the horizon, greatly affecting performance particularly towards the horizon where effects of the lower layers of atmosphere are amplified manyfold.

One approximate model for solar intensity versus airmass is given by:

$$I = 1.1 \times I_o \times 0.7^{(AM^{0.678})}$$

where solar intensity external to the Earth's atmosphere I_o = 1.353 kW/m2, and the factor of 1.1 is derived assuming that the diffuse component is 10% of the direct component.

This formula fits comfortably within the mid-range of the expected pollution-based variability:

Solar intensity vs. zenith angle z and airmass coefficient AM				
z	AM	range due to pollution	formula (I.1)	ASTM G-173
degree		W/m²	W/m²	W/m²
-	0	1367	1353	1347.9
0°	1	840 .. 1130 = 990 ± 15%	1040	
23°	1.09	800 .. 1110 = 960 ± 16%	1020	
30°	1.15	780 .. 1100 = 940 ± 17%	1010	
45°	1.41	710 .. 1060 = 880 ± 20%	950	
48.2°	1.5	680 .. 1050 = 870 ± 21%	930	1000.4
60°	2	560 .. 970 = 770 ± 27%	840	
70°	2.9	430 .. 880 = 650 ± 34%	710	
75°	3.8	330 .. 800 = 560 ± 41%	620	
80°	5.6	200 .. 660 = 430 ± 53%	470	
85°	10	85 .. 480 = 280 ± 70%	270	
90°	38		20	

This illustrates that significant power is available at only a few degrees above the horizon. For example, when the sun is more than about 60° above the horizon (z <30°) the solar intensity is about 1000 W/m², whereas when the sun is only 15° above the horizon (z =75°) the solar intensity is still about 600 W/m² or 60% of its maximum level; and at only 5° above the horizon still 27% of the maximum.

At Higher Altitudes

One approximate model for intensity increase with altitude and accurate to a few kilometres above sea level is given by:

$$I = 1.1 \times I_o \times [(1 - h/7.1)0.7^{(AM)^{0.678}} + h/7.1]$$

where h is the solar collector's height above sea level in km and AM is the airmass
(from A.2) as if the collector was installed at sea level.

Alternatively, given the significant practical variabilities involved, the homogeneous
spherical model could be applied to estimate AM, using:

$$AM = \sqrt{(r+c)^2 \cos^2 z + (2r+1+c)(1-c)} - (r+c)\cos z$$

where the normalized heights of the atmosphere and of the collector are respectively
$r = R_E / y_{atm} \approx 708$ (as above) and $c = h/y_{atm}$.

And then the above table or the appropriate equation (I.1 or I.3 or I.4 for average, polluted or clean air respectively) can be used to estimate intensity from AM in the normal way.

These approximations at I.2 and A.4 are suitable for use only to altitudes of a few kilometres above sea level, implying as they do reduction to AM0 performance levels at only
around 6 and 9 km respectively. By contrast much of the attenuation of the high energy
components occurs in the ozone layer - at higher altitudes around 30 km. Hence these approximations are suitable only for estimating the performance of ground-based collectors.

Solar Cell Efficiency

Silicon solar cells are not very sensitive to the portions of the spectrum lost in the atmosphere. The resulting spectrum at the Earth's surface more closely matches the band-
gap of silicon so silicon solar cells are more efficient at AM1 than AM0. This apparent-
ly counter-intuitive result arises simply because silicon cells can't make much use of
the high energy radiation which the atmosphere filters out. As illustrated below, even
though the *efficiency* is lower at AM0 the *total output power* (P_{out}) for a typical solar
cell is still highest at AM0. Conversely, the shape of the spectrum does not significantly
change with further increases in atmospheric thickness, and hence cell efficiency does
not greatly change for AM numbers above 1.

Output power vs. airmass coefficient			
AM	Solar intensity P_{in} W/m²	Output power P_{out} W/m²	Efficiency P_{out} / P_{in}
0	1350	160	12%
1	1000	150	15%
2	800	120	15%

This illustrates the more general point that given that solar energy is "free", and where available space is not a limitation, other factors such as total P_{out} and $P_{out}/\$$ are often more important considerations than efficiency (P_{out}/P_{in}).

THE EFFECT OF AIR MASS ON SOLAR PANELS

As the Sun's rays pass through the atmosphere, the energy of some photons of light is absorbed and scattered by particles in the atmosphere. The lower the Sun is in the sky (closer to the horizon), the thicker the layer of atmosphere the Sun's rays must pass through to get to the solar panel down here on the Earth's surface, and therefore the more energy is absorbed in the atmosphere which is therefore not available for the solar panel to convert into electricity.

The distance the Sun's rays have to pass through the atmosphere can make a significant reduction in the amount of electricity which can be generated by a PV solar panel or the amount of hot water which can be made by a solar water heating system, particularly in higher latitudes such as here in the UK and especially in the winter months when the Sun is low in the sky all day.

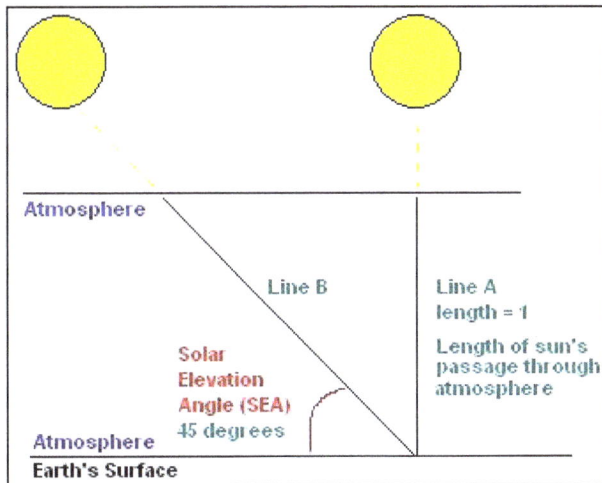

The above diagram illustrates this effect. Line A shows the Sun directly overhead. If the thickness of the atmosphere through which the Sun's rays pass is given by 1 unit of length, we can work out the length of line B which shows the Sun at a solar elevation angle (SEA) of 45° above the horizon:

Line B = 1 / sin (45) = 1.41

Therefore when the Sun is at an SEA of 45 degrees, the Sun's rays pass through 41% more atmosphere than they would if the Sun were overhead (SEA = 90).

Air Mass and Solar Insolation/Intensity

Solar Insolation is a measure of the amount of incident solar radiation (the solar intensity) directly hitting a given part of the Earth's surface. It has to be measured rather than calculated since it depends on latitude, humidity, air pollution and smog, local weather, altitude, and even the shading effects of local mountains.

Solar intensity is measured in W/m² (Watts per square metre), and at AM1 (when the Sun is directly overhead) it is approximately 1040 W/m² at sea level.

As the Sun moves down toward the horizon, the air mass value increases as shown earlier, and the solar intensity level falls – i.e. as the Sun moves lower in the sky, the amount of solar energy getting through the atmosphere and to ground level goes down as more energy is absorbed and scattered by particles in the atmosphere.

The relationship between air mass value and solar intensity has been modelled (experimentally determined) and the following equation gives a good approximation at sea level:

$$I = 1.1 \times I_o \times 0.7^{(AM)^{(0.678)}}$$

where AM is the air mass, Io is the intensity of the Sun hitting the top of the atmosphere (1.353 kW/m²), and I is the intensity of the Sun reaching a flat surface perpendicular to the Sun's rays at sea level.

Trying different air mass values with this equation, it becomes clear that the effect of air mass on solar insolation is not significant when the sun is high in the sky. With AM1.15 (solar elevation angle 60°) the intensity is 1010 W/m², less than a 3% drop from AM1 (Sun overhead). With AM1.41 (solar elevation angle 45°) the intensity is still 950 W/m2, only a 9% drop from AM1.

It is only when we get down below around 30° solar elevation angle that the effect of air mass becomes really significant, for example when the sun has moved down to just over 10° above the horizon, half of the energy of the Sun's rays are lost on their way through the atmosphere.

SOLAR ENERGY POWER GENERATION

Solar energy generation is one of fastest growing and most promising renewable energy sources of power generation worldwide. Nowadays, the electrical energy becomes one of the basic needs in our daily life, which makes increasing demand for it.

As a major source of electrical power generation fossil fuels are depleting day by day and also its usage raises serious environmental concerns. These reasons force the development of new energy sources which are renewable and ecologically safe.

Solar Energry

The renewable energy sources include wind, solar, water, biomass and geothermal energy sources. Out of which, solar energy has the greatest potential in the long term and is predicted to play a major role in coming years. It is the cheapest method of generating electricity compared with other energy sources.

How Solar Power Generated from PV Cell?

A PV cell (can be called as a solar cell) is a semiconductor device that converts the sunlight energy into electricity without going through any energy conversion steps.

This conversion takes place by photovoltaic effect and hence they are called Photovoltaic (PV) cells. It generates voltage and current at its terminals when sunlight incident on it.

PV Cell.

The way and the amount of power generated by a solar cell depend on the sunlight falling on it. This also includes some factors such as intensity of light, angle at which the light falls on it and area of the cell.

The more is the power generated, if higher is the light intensity. If the area of the cell is more, the power generated is also more. And the optimum power is generated by it when light falling is perpendicular to the front side of the cell.

The solar cells are made with silicon semiconductor material and is treated with phosphorous and boron to make a thin silicon wafer. The wafer layers are then aligned together to make the solar cells, once they are doped.

Irrespective of the technology and material used, every solar cell has two terminals (positive and negative terminals) so as to take the electric current from it. Typically, a solar cell consists of front contact at the top, PN junction in the middle and back contact at the bottom.

Basically, the sunlight consists of bundles of photons, where each photon has a finite amount of energy. To generate the electricity from a solar cell, these photons must be absorbed by it. The energy of the photon and also the band-gap energy of semiconductor material decide the absorption of a photon.

Here is the term Electron-volt (eV) which is the unit of energy that expresses the photon energy and the band-gap energy of a semiconductor material.

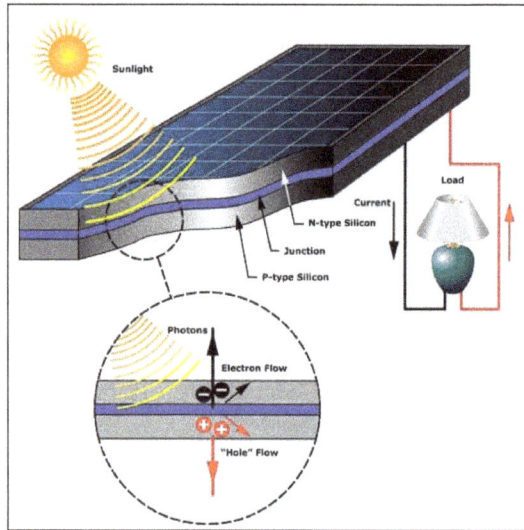
Photo electric effect.

The semiconductor material of the solar panel absorbs the photons in the sunlight. Due to this, electron-hole pairs are generated at the junction. When the solar cell is connected to the load, electrons and holes at the junction are separated from each other where the electrons are collected at the negative terminal and holes at positive terminal.

Thus the electric potential is built between the terminals and hence the voltage is developed across it. This further drives the current (DC) to the DC loads, inverter, or battery charging circuit. If more photons are absorbed, greater will be the current generated. However, much of the solar radiation fall on the solar cell is not converted into electricity.

This is because light is composed of photons of different wavelengths. Some photons hit the solar cell and then reflected and prevented from entering to the cell. In some materials, generated electrons recombine with other molecules before being drawn into current. Likewise, there are many reasons for the low conversion rate or efficiency. The conversion efficiency of the solar panels used in individual residences ranges from 6 to 10%.

And for large-scale installations and solar power plants, solar panels are designed with best material and technologies to achieve higher conversion efficiency ranging from 40 to 60%, but they are costliest.

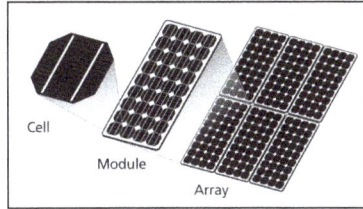

A single solar cell of 4 cm² develops a voltage of 0.5 to 1 V and it can produce 0.7W power when exposed to the sunlight. Typically, the best designed solar panel has a maximum efficiency of 25%.

In order to generate high potential difference or voltage and more electric power, these individual cells are connected together that means some cells are connected in series and some are in parallel.

PV modules are formed by connecting the number of number of solar cells together. And several PV modules are connected together to make a PV array which can be used for small power as well as high power generation applications.

Components of Solar Energy Electricity Generation

The major components in solar energy electricity generation include:

Solar Panels

Solar cell modules or solar panels convert the solar energy into electricity. These are mounted in such a way that they collect maximum energy from the sun. Most solar panels are rated to a voltage 12V (a half volt PV cells are connected in series, inside of the solar panel to produce the high voltage say 12V).

The panels are connected in series to form a solar array that produces higher voltage, typically of 24 or 48V in standalone systems, or it can be several hundreds of volts in grid-connected systems.

If the panels are connected in parallel, the current delivered to the load will be more and hence the more power while maintaining the same voltage. Irrespective of series or parallel connection, the power rating of the system increases when multiple solar panels are connected together.

Solar Panels.

Here, a solar array is made with four 12V and 12 watt solar panels, where each panel produces a current of 1 A. Then this array would be rated as 48V, 48W with 1A current. If the same rated solar panels are connected in parallel to form an array of four solar panels, then the solar array would be rated as 12V, 48W with 4 A current.

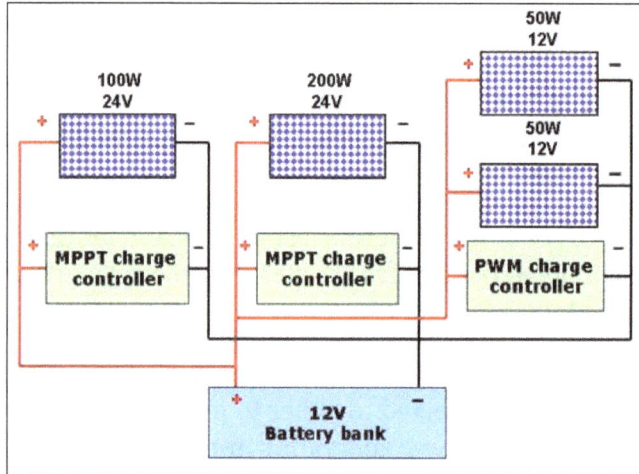

Batteries

Batteries.

Except the grid-connected system, all other solar energy power generation systems use batteries to store the energy generated from solar panels. Since the amount of solar power generated depends on the strength of the sunlight, batteries provide a constant source of power supply once it is fully charged.

Mostly, lead acid batteries are used in solar electric systems. Like solar panels, batteries can be connected together to form a battery bank.

These can be connected either series or parallel as similar to solar panels to achieve desired voltage, current and power ratings. The type of battery chosen depends on the energy requirements of a system and its budget.

Controller

It regulates the flow of current into and out of the battery. If the generated current overcharges the battery, it leads to damage in the battery. Moreover, if the battery is completely discharged, it will destroy the battery. Hence the solar controller prevents the batteries to undergo these conditions.

Solar Charge Controller.

The charge controller module balances the amount of electricity used to power appliance and lights with generated power. Also, it prevents the damage to the batteries due to overcharging and deep discharging. In addition, it gives the alarm when the module not functioning properly.

Inverter

The electricity generated from the solar panels is a Direct Current (DC), whereas the most electrical appliances work on Alternating Current (AC) and hence a converter is needed to convert DC to AC, nothing but an inverter.

Also, if the solar system is connected to the grid, the generated DC voltage must be converted into AC. So the inverter equipment converts the DC voltage to the AC and to the same voltage as that of grid or appliance rating.

As a recent invention, most individual solar panels are connected with micro inverters that provide a high AC voltage. These are suitable only for grid-connected systems and not suitable with battery backup systems.

Inverter.

Different Types of Solar Electric Systems

Off-Grid/Standalone Solar Electric Systems

These are the most popular type of solar installations which are primarily designed to supplement or replace the conventional mains supply.

These are mainly used in the locations where there is no other source to provide power supply and hence these are used in remote locations and rural areas where it is difficult to get the power from grid extensions.

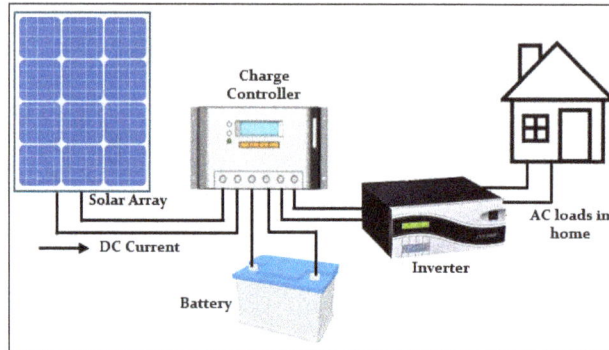

Generally, off-grid systems use solar power to charge the batteries, and this charge is then supplied to the load when needed. The battery power either directly operates the DC loads (DC lamps) or drives the power inverter that converts the DC power to AC power to operate the appliance like pumps, lighting equipment, refrigerators, etc.

This method is followed for any standalone system whether it is a pocket calculator or a complete off-grid home. Standalone systems are comparatively small and simple systems.

Grid-Connected Solar Electric Systems

These systems effectively create a micro power station and are connected directly into the electricity grid. These are normally found in urban areas where power is readily available.

During the day it feeds the excess electricity generated into the grid and during evening and night it imports the electricity from the grid. Here the grid acts like a storage medium in which power is taken from the grid when needed.

It is to be noted that, grid-connected system doesn't have to supply enough electricity to cover entire power demand. So this system can be small or large depends on the owner's choice.

This system receives the payment for each kilowatt of power which is supplied to the electricity providers. This type of installation reduces the dependence on electric utilities and hence reduces the electricity bills.

The major components in this solar installation include a PV array or solar cells, inverter and the metering system. The main disadvantage of the grid-tied solar system is that it will switch off in the event of power cut, because this system is the part of utility grid and hence if there is a power cut, the power from the solar array is also switch off.

If this system won't stop, current flow back into the grid could lead to serious faults.

Central Grid-Connected Solar Electric Systems

In a conventional grid system a variety of power sources such as gas, coal, water, etc. combined and it supplies the power to end user via transmission and distribution lines.

Likewise, a central grid connected system is directly connected to the transmission lines. These systems can be small (as 50kWp), large (as 100kWp) or somewhat higher range say 1GWp which all are directly connected to the transmission systems. Mostly, these are called as solar power plants.

Grid Fallback Solar Electric Systems

It combines the grid connected system with a bank of batteries. In this solar array generates the electricity, which in turn charges a battery bank. The battery power is then running the inverter which drives the loads through an inverter.

When the power in the batteries is not enough to drive the loads, the system automatically switches back to the grid power supply. Again, it switches back to battery power once the solar array recharges the batteries.

This system doesn't sell any power to any electrical utilities and the overall power generated used for domestic or residential systems alone.

References

- Solar-energy, science: britannica.Com, retrieved 8 january, 2019

- "Indirect gain (trombe walls)". United states department of energy. Archived from the original on 15 april 2012. Retrieved 2007-09-29

- Kraemer, d; hu, l; muto, a; chen, x; chen, g; chiesa, m (2008), "photovoltaic-thermoelectric hybrid systems: a general optimization methodology", applied physics letters, 92 (24): 243503, bibcode:2008apphl..92X3503k, doi:10.1063/1.2947591

- Solar-energy-power-generation: electronicshub.Org, retrieved 13 may, 2019

- Chronicle of fraunhofer-gesellschaft". Fraunhofer-gesellschaft. Archived from the original on 12 december 2007. Retrieved 4 november 2007

- Sanjari, m.J.; Gooi, h.B. (2016). "Probabilistic forecast of pv power generation based on higher-order markov chain". Ieee transactions on power systems. Doi:10.1109/Tpwrs.2016.2616902

- The-effect-of-air-mass-on-solar-panels, solar, wordpress: reuk.Co.Uk, retrieved 25 february, 2019

- Perry, keith (28 july 2014). "Most solar panels are facing the wrong direction, say scientists". The daily telegraph. Retrieved 9 september 2018

3

Understanding Wind Energy

The utilization of air flow with the help of wind turbines to provide mechanical power is referred to as wind energy. The production of power and electricity through wind energy takes place at wind farms. This chapter has been carefully written to provide an easy understanding of the varied facets of wind energy as well as its advantages and disadvantages.

Wind Energy is the most mature and developed renewable energy Wind energy (or wind power) describes the process by which wind is used to generate electricity. Wind turbines convert the kinetic energy in the wind into mechanical power. A generator can convert mechanical power into electricity. Mechanical power can also be utilized directly for specific tasks such as pumping water. The US DOE developed a short wind power animation that provides an overview of how a wind turbine works and describes the wind resources in the United States.

Wind is caused by the uneven heating of the atmosphere by the sun, variations in the earth's surface, and rotation of the earth. Mountains, bodies of water, and vegetation all influence wind flow patterns. Wind turbines convert the energy in wind to electricity by rotating propeller-like blades around a rotor. The rotor turns the drive shaft, which turns an electric generator. Three key factors affect the amount of energy a turbine can harness from the wind: wind speed, air density, and swept area.

Equation for Wind Power

$$P = \frac{1}{2}\rho A V^3$$

- Wind speed

  The amount of energy in the wind varies with the cube of the wind speed, in other words, if the wind speed doubles, there is eight times more energy in the wind $(2^3 = 2x2x2 = 8)$. Small changes in wind speed have a large impact on the amount of power available in the wind.

- Density of the air

  The more dense the air, the more energy received by the turbine. Air density varies with elevation and temperature. Air is less dense at higher elevations than at sea level, and warm air is less dense than cold air. All else being equal, turbines will produce more power at lower elevations and in locations with cooler average temperatures.

- Swept area of the turbine

  The larger the swept area (the size of the area through which the rotor spins), the more power the turbine can capture from the wind. Since swept area is $A = pir^2$, where r = radius of the rotor, a small increase in blade length results in a larger increase in the power available to the turbine.

Wind Farm Development

Siting a wind farm varies from one location to another, but there are some important matters for land owners to consider:

- Understand your wind resource.

- Evaluate distance from existing transmission lines.

- Determine benefits of and barriers to allowing your land to be developed.

- Establish access to capital.

- Identify reliable power purchaser or market.

- Address siting and project feasibility considerations.

- Understand wind energy's economics.

- Obtain zoning and permitting expertise.

- Establish dialogue with turbine manufacturers and project developers.

- Secure agreement to meet O&M needs.

Land Requirements

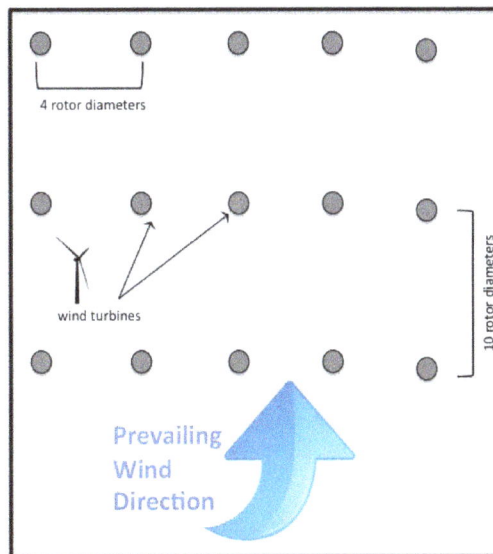

The amount of land required for a wind farm varies considerably, and is particularly dependent on two key factors: the desired size of the wind farm (which can be defined either by installed capacity or the number of turbines) and the characteristics of the local terrain. Typically, wind turbine spacing is determined by the rotor diameter and local wind conditions. Some estimates suggest spacing turbines between 5 and 10 rotor diameters apart. If prevailing winds are generally from the same direction, turbines may be installed 3 or 4 rotor diameters apart (in the direction perpendicular to the prevailing winds); under multi-directional wind conditions, spacing of between 5 and 7 rotor diameters is recommended.

Advantages and Disadvantages of Wind Energy

Advantages of Wind Energy

Wind energy has a number of different benefits. We can use it for a variety of purposes, primarily for the production of clean and renewable electricity.

Wind Energy is Renewable and Sustainable

Wind energy itself is both renewable and sustainable. The wind will never run out, unlike reserves of fossil fuels (such as coal, oil, and gas.) This makes it a good choice of energy for a sustainable power supply.

It's also Environmentally Friendly

Wind energy is one of the most environmentally friendly energy sources available today. This is based on the simple reason that wind turbines don't create pollution when generating electricity.

Most non-renewable energy sources need to be burnt. This process releases gases such as carbon dioxide (CO_2) and methane (CH_4) into the atmosphere. These gases are known to contribute to climate change. In contrast, wind turbines produce no greenhouse gases when generating electricity.

We should note that both noise and visual pollution are environmental disadvantages of wind turbines. However, these factors don't have a negative impact on the earth, water table or the quality of the air we breathe.

It can Reduce Fossil Fuel Consumption

Generating electricity from wind energy reduces the need to burn fossil fuel alternatives such as coal, oil, and gas. This can help to conserve dwindling supplies of the earth's natural resources. As a result, they will last longer and help to support future generations.

Wind Energy is Free

Unlike most non-renewable energy sources, wind energy is completely free. Anyone can make use of the wind and it will never run out. This makes wind energy a viable option for generating cheap electricity.

It has a Small Footprint

Land surrounding wind turbines can be used for other purposes such as agriculture.

Wind turbines have a relatively small land footprint. Although they can tower high above the ground, the impact on the land at the base is minimal. Wind turbines are often constructed in fields, on hills or out at sea. At these locations, they pose hardly any inconvenience to the surrounding land. Farmers can still farm their fields, livestock can still graze the hills and fishermen can still fish the sea.

Both Industrial and Domestic Wind Turbines Exist

Wind turbines aren't just limited to industrial-scale installations (such as wind farms.) They can also be installed on a domestic scale. As a result, many landowners opt to install smaller, less powerful wind turbines. This can help to provide a portion of a domestic electricity supply. Domestic wind turbines are often coupled with other renewable energy technologies. You can often find them installed alongside solar panels and geothermal heating systems.

Wind Energy can Provide Power for Remote Locations

Wind turbines can play a key role in helping to bring power to remote locations. This can help to benefit everything from small off-grid villages to remote research facilities. It might be impractical or too expensive to hook such locations up to traditional electricity supplies. In these cases, wind turbines could have the answer.

Wind turbines can be used to generate power in remote locations.

Wind Technology is Becoming Cheaper

The first-ever wind turbine started generating electricity in 1888. Since then, they have become more efficient and have come down in price. As a result of this, wind power is becoming much more accessible.

Government subsidies are also helping to reduce the cost of wind technologies. Many countries across the world now provide incentives for the construction of wind turbines. In addition, incentives are sometimes available for domestic users to supply electricity back to the grid.

It is also Low Maintenance

Wind turbines are fairly low in maintenance. A new wind turbine can last a long time prior to it requiring any maintenance. Although older turbines can come up against reliability issues, technological advancements are helping to improve overall reliability.

It has Low Running Costs

As wind energy is free, running costs are often low. The only ongoing cost of wind energy is for the maintenance of wind turbines, but they are low maintenance in nature anyway.

Wind Energy has Huge Potential

Wind energy has huge potential. It's both renewable and sustainable and is present in a wide variety of places. Although wind turbines aren't cost-effective at every location, the technology isn't limited to just a handful of locations. This is an issue that can affect other renewable energy technologies – such as geothermal power stations.

It can Increase Energy Security

By using wind energy to generate electricity, we are helping to reduce our dependency on fossil fuel alternatives. In many cases, a country will source some or all of its fossil fuels from another country. War, politics and overall demand often dictate the price of these natural resources. This can sometimes cause serious economic problems or supply shortages.

By using local renewable energy sources, a country can reduce its dependency on external supplies of natural resources. As a direct result of this, the country can increase its energy security.

The Wind Energy Industry Creates Jobs

The wind energy industry has boomed since wind turbines became commercially available. As a result of this, the industry has created jobs all over the world. Jobs now exist for the manufacturing, installation, and maintenance of wind turbines. You can even find jobs in wind energy consulting. This is a job where specialist consultants determine whether a wind turbine installation is going to be profitable.

The renewable energy industry employed over 10 million people worldwide in 2017. Of these jobs, 1.15 million were in the wind power industry. China leads the way in providing over 500,000 of these jobs. Germany is in second place with around 150,000 jobs and the United States are a close third with around 100,000 wind energy jobs.

Disadvantages of Wind Energy

Wind energy has a number of drawbacks and cons.

The Wind Fluctuates

Wind energy has a similar drawback to solar energy in that it is not constant. Although wind energy is sustainable and will never run out, the wind isn't always blowing. This can cause serious problems for wind farm developers. They will often spend a significant amount of time and money investigating whether a particular site is suitable for wind power.

For a wind turbine to be efficient, it needs to have an adequate supply of wind energy. For this reason, we often find wind turbines on top of hills or out at sea. In these locations, there are fewer land obstacles to reduce the force of the wind.

Installation is Expensive

Although costs are reducing over time, wind turbines are still expensive. First, an engineer must carry out a site survey. This may involve having to erect a sample turbine to measure wind speeds over a period of time. If deemed adequate, a wind turbine then needs to be manufactured, transported and erected on top of a pre-built foundation. All of these processes contribute to the overall cost of installing wind turbines.

When we take the above into account for offshore wind farms, the costs become much greater. Installing structures out at sea is far more complex than on land. Some companies have even commissioned bespoke ships capable of transporting and installing wind turbines at sea.

Installing wind turbines is an expensive process.

Wind Turbines Pose a Threat to Wildlife

We often hear that wind turbines pose a threat to wildlife – primarily birds and bats. However, researchers now believe that they pose less of a threat to wildlife than other manmade structures. Installations such as cell phone masts and radio towers are far more dangerous to birds than wind turbines. Nevertheless, wind turbines still contribute to mortality rates among bird and bat populations.

Wind Turbines Create Noise Pollution

One of the most common disadvantages of wind turbines is the noise pollution they generate. You can often hear a single wind turbine from hundreds of meters away. Combine multiple wind turbines with the right wind direction and the audible effects can be much greater. This issue is one of the biggest impacts of wind energy.

Noise pollution from wind turbines has ruined the lives of many homeowners. Although steps are often taken to install them away from dwellings, they do sometimes get built too close to where people live. This is why new wind farms often come up against strong public objection.

They also Create Visual Pollution

Another common drawback of wind turbines is the visual pollution they create. Although many people actually like the look of wind turbines, others don't. These people see them as a blot on the landscape. This, however, tends to come down to personal opinion. As we build more wind farms, public acceptance is becoming more common.

Some people see wind turbines as 'visual pollution'.

WIND TURBINES

Wind turbines are the evolution of the classic windmills that can be seen in more rural areas of the world. Their purpose is to reduce reliance on fossil fuels to create energy and also to create energy in a less wasteful manner. They operate by using the kinetic energy of the wind, which pushes the blades of the turbine and spins a motor that converts the kinetic energy into electrical energy for consumer use.

Wind Turbines are rotating machines that can be used directly for grinding or can be used to generate electricity from the kinetic power of the wind. They provide the clean and renewable energy for us of both home and office. Wind Turbines are a great way to save money and make the environment clean and green.

This process has been adapted for use for various applications and can be seen in use by boats, traffic signs, or whole communities that use a wind farm for power. The development of wind turbines is a major step towards overhauling the way we produce our energy.

Basically there are two types of wind generators, those with vertical axis and those with horizontal axis. They can be used to generate electricity both onshore and offshore. Wind Turbines can be combined to form clusters called "wind farms" which are used by large companies to use that power as their backup. Apart from generating electricity they can also be used for grain-grinding, water pumping, charging batteries.

Historically, wind turbines were used for sailing, irrigation and grinding-grains. It was in the early 20th century that it was used for generating power. Today, large wind turbines can be seen in the rural areas or near the sea coast where speed of the wind is generally throughout the day. Device called wind resource assessment is used for estimating the wind speed.

Wind Turbine Components

Wind turbine systems are made up of many different pieces of equipment that all serve a purpose in delivering electricity where it is intended to go. Of those many different pieces, the below list serves as a general blueprint for the main components that can and often times will be found in wind turbine systems regardless of the type of design.

- Rotor: The rotor is made of blades that are attached to a center piece. The blades are shaped such that when the wind pushes against them they turn.

- Pitch Drive: Used to rotate blades to accommodate for high-speed wind.

- Nacelle: The rotor is attached to a housing unit called a nacelle, which protects various other components necessary to the wind turbine operation.

- Brake: Necessary to slow the rotor down.

- Low-Speed Shaft: Attaches to the rotor and turns as the rotor turns on a 1:1 ratio.

- Gear Box: Serves the same function as a car, the rotor spins slowly as the wind pushes against it and the gearbox or transmission increases that rotational speed for the generator.

- High-Speed Shaft: Attaches to the gearbox and generator and spins at a higher speed than the rotor or low-speed shaft.

- Generator: Actual mechanism that converts the rotational kinetic energy into electricity.

- Wind Vane: Detects direction of wind and adjusts the rotor and nacelle to compensate.

- Yaw Drive: Keeps the rotor and therefore the turbines facing the wind.

- Tower: Elevates the aforementioned components to an altitude that optimizes wind exposure.

How do Wind Turbines Work?

Wind Turbines operate on a system comprised of many critical components that allow kinetic wind energy to be transformed into electrical energy. No matter the type of wind turbine system, they all work off the same principle that allows a generator to produce electricity. This principle is that if magnets are rotated around a coil of wire, or a coil of wire rotates within a magnetic field fast enough electricity is produced.

The total contraption of magnets and conductors make up the generator. Using the wind to turn blades creates the force needed to turn the magnets or the coil of conductor which in turn creates electricity. Below is a step by step process that highlights the methods by which a wind turbine actually comes to produce electricity.

- Tower is constructed that puts the wind turbine system at the correct altitude where wind travels at a higher and more constant rate.

- Rotor blades are exposed to wind which forces them to start turning.

- As the rotor spins, the low-speed shaft, which is connected to a gearbox, spins at the same rate.

- The gearbox takes this slow rotational speed and through correct gearing turns it into a faster rotational speed.

- The high-speed shaft, which is on the outgoing end of the gearbox and connected to a generator, spins at a higher rate of speed.

- The generator spins at this high rate of speed which spins magnets around a coil of metal wire and generates electricity.

- The electricity travels from the generator through wires to the necessary applications whether it be direct appliances or a battery.

Types of Wind Turbines

There are two main types of wind turbines that can be seen in design and implementation in the wind energy industry today. The first and most common type is the horizontal axis wind turbine that relies on a horizontal shaft that runs perpendicular to the blades which spin vertically. These wind turbine systems can be seen in use in major wind farms as well as solo operations.

The second type which is less common among the wind energy industry is the vertical axis wind turbine. As one might be able to infer, the vertical axis turbine has a vertical shaft in which the blades or rotor are connected to and spin horizontally. There are many variations of the vertical axis wind turbine but the major benefit is that maintenance is easier because the gearbox and generator are more accessible.

- Horizontal Axis Wind Turbine: This is the standard type of wind turbine where the low-speed shaft that connects to the rotor is horizontal. There are various ways to construct this wind turbine but they all follow the same concept as outlined above. The rotor spins with the wind and the rotational kinetic energy is converted to electrical energy through a generator.

- Vertical Axis Wind Turbine: This type of wind turbine is less common but has an advantage in that the rotor does not need to face into the wind. The shaft connecting to the rotor is vertical and the gearbox and generator are generally at the bottom of the tower. There are many types of vertical axis wind turbines all of which follow the same concept of force along the X-axis (parallel to the ground) as opposed to horizontal axis turbines which use force along the Y-axis (perpendicular to the ground).

The following are different variations that come from vertical axis wind turbine systems. Many of these were engineered decades ago and are no longer seen in use today, however the designs for these have been adapted and tweaked such that newer models can be developed that are more efficient with less problems than the older ones.

- Darrieus Wind Turbine: This vertical axis wind turbine uses curved blades that rotate and creates an internal force of wind that enables the rotor to spin at high speeds regardless of the wind speed. The downside of this is that this turbine generally requires an external motor to start spinning.

- Giromill: A variation of the Darrieus Wind Turbine in that it uses an H shaped rotor. The difference between the two is that the giromill uses straight blades that run parallel to the shaft. Other than that, the two operate on the entirely same principal.

- Cycloturbine: A type of giromill that not only has straight blades running vertically but also that the straight blade can itself rotate around its center axis. The advantage in this type of turbine is that it generates the most amount of power and can self-start (start without any external assistance).

- Savonius Wind Turbine: This vertical axis wind turbine relies on the principles of drag and wind resistance to function. The blades are shaped like an S with the two curved parts of the S moving with the wind. The curved part creates less drag and therefore the rotor is able to spin. These turbines do not generate much energy.

- Vortexis Wind Turbine: This is the most recent development of vertical axis wind turbines. It has seen use in Afghanistan and Iraq by special forces needing to power their devices. This turbine has two sets of blades, one smaller set that sits in a circle and one bigger set that surrounds the smaller set in a larger circle, that act in a gearbox like fashion. The outer set of blades use the wind to spin and by that set of blades spinning they force their own wind to turn the smaller inside set of blades. These blades are connected to the shaft which then turns a generator.

WIND FARMS

A wind farm or wind park, also called a wind power station or wind power plant, is a group of wind turbines in the same location used to produce electricity. A large wind farm may consist of several hundred individual wind turbines and cover an extended area of hundreds of square miles, but the land between the turbines may be used for agricultural or other purposes. A wind farm can also be located offshore.

The San Gorgonio Pass wind farm in California, United States.

Many of the largest operational onshore wind farms are located in China, India, and the United States. For example, the largest wind farm in the world, Gansu Wind Farm in China had a capacity of over 6,000 MW by 2012, with a goal of 20,000 MW by 2020. As of September 2018, the 659 MW Walney Wind Farm in the UK is the largest offshore wind farm in the world. Individual wind turbine designs continue to increase in power, resulting in fewer turbines being needed for the same total output.

Wind farms tend to have much less impact on the environment than many other power stations. Onshore wind farms are also criticized for their visual impact and impact on the landscape, as typically they need to take up more land than other power stations and need to be built in wild and rural areas, which can lead to "industrialization of the countryside", habitat loss, and a drop in tourism. Critics have linked wind farms to adverse health effects. Wind farms have also been criticized for interfering with radar, radio and television reception.

Design and Location

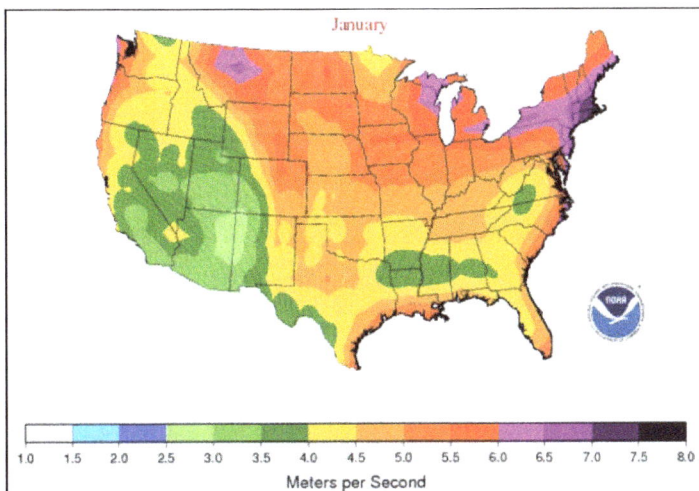

Map of available wind power over the United States. Color codes indicate wind power density class.

The location is critical to the success of a wind farm. Conditions contributing to a successful wind farm location include: wind conditions, access to electric transmission, physical access, and local electric prices.

The faster the average wind speed, the more electricity the wind turbine will generate, so faster winds are generally economically better for wind farm developments. The balancing factor is that strong gusts and high turbulence require stronger more expensive turbines, otherwise they risk damage. The average power in the wind is not proportional to the average wind speed, however. For this reason, the ideal wind conditions would be strong but consistent winds with low turbulence coming from a single direction.

Mountain passes are ideal locations for wind farms under these conditions. Mountain passes channel wind blocked by mountains through a tunnel like pass towards areas of lower pressure and flatter land. Passes used for wind farms like the San Gorgonio Pass and Altamont Pass are known for their abundant wind resource capacity and capability for large-scale wind farms. These types of passes were the first places in the 1980's to have heavily invested large-scale wind farms after approval for wind energy development by the U.S. Bureau of Land Management. From these wind farms, developers learned a lot about turbulence and crowding effects of large-scale wind projects previously unresearched in the U.S. due to the lack of operational wind farms large enough to conduct these types of studies on.

Usually sites are screened on the basis of a wind atlas, and validated with on-site wind measurements via long term or permanent meterological-tower data using anemometers and wind vanes. Meteorological wind data alone is usually not sufficient for accurate siting of a large wind power project. Collection of site specific data for wind speed and direction is crucial to determining site potential in order to finance the project. Local winds are often monitored for a year or more, detailed wind maps are constructed, along with rigorous grid capability studies conducted, before any wind generators are installed.

Part of the Biglow Canyon Wind Farm, Oregon, United States
with a turbine under construction.

The wind blows faster at higher altitudes because of the reduced influence of drag. The increase in velocity with altitude is most dramatic near the surface and is affected by topography, surface roughness, and upwind obstacles such as trees or buildings. However, at higher altitudes, the power in the wind decreases proportional to the decrease in air density. Rendering significantly less efficient power extraction by the wind turbines, requiring for a higher investment for the same generation capacity at lower altitudes.

How closely to space the turbines together is a major factor in wind farm design. The closer the turbines are together the more the upwind turbines block wind from their neighbors. However spacing turbines far apart increases the costs of roads and cables, and raises the amount of land needed to install a specific capacity of turbines. As a result of these factors, turbine spacing varies by site. Generally speaking manufacturers require 3.5 times the rotor diameter of the turbine between turbines as a minimum. Closer spacing is possible depending on the turbine model, the conditions at the site, and how the site will be operated.

Often in heavily saturated energy markets, the first step in site selection for large-scale wind projects before wind resource data collection is finding areas with adequate Available Transfer Capability (ATC). ATC is the measure of the remaining capacity in a transmission system available for further integration of generation without significant upgrades to transmission lines and substations, which have substantial costs, potentially undermining the viability of a project within that area, regardless of wind resource availability. Once a list of capable areas is constructed, the list is refined based on long term wind measurements, among other environmental or technical limiting factors such as proximity to load and land procurement.

Many Independent System Operators (ISO's) in the United States such as the California ISO and Midcontinent ISO use interconnection request queues to allow developers to propose new generation for a specific given area and grid interconnection. These request queues have both deposit costs at the time of request and ongoing costs for the studies the ISO will make for up to years after the request was submitted to ascertain the viability of the interconnection due to factors such as ATC. Larger corporations who can afford to bid the most queues will most likely have market power as to which sites with the most resource and opportunity get to be developed upon. After the deadline to request a place in the queue has passed, many firms will withdraw their requests after gauging the competition in order to make back some of the deposit for each request that is determined too risky in comparison to other larger firms' requests.

Onshore

An aerial view of Whitelee Wind Farm, the largest onshore wind farm in the UK and second-largest in Europe.

The world's first wind farm was 0.6 MW, consisting of 20 wind turbines rated at 30 kilowatts each, installed on the shoulder of Crotched Mountain in southern New Hampshire in December 1980.

Onshore turbine installations in hilly or mountainous regions tend to be on ridges generally three kilometres or more inland from the nearest shoreline. This is done to exploit the topographic acceleration as the wind accelerates over a ridge. The additional wind speeds gained in this way can increase energy produced because more wind goes through the turbines. The exact position of each turbine matters, because a difference of 30m could potentially double output. This careful placement is referred to as 'micro-siting'.

Offshore

Offshore wind turbines near Copenhagen, Denmark.

Europe is the leader in offshore wind energy, with the first offshore wind farm (Vindeby) being installed in Denmark in 1991. As of 2010, there are 39 offshore wind farms in waters off Belgium, Denmark, Finland, Germany, Ireland, the Netherlands, Norway, Sweden and the United Kingdom, with a combined operating capacity of 2,396 MW. More than 100 GW (or 100,000 MW) of offshore projects are proposed or under development in Europe. The European Wind Energy Association has set a target of 40 GW installed by 2020 and 150 GW by 2030.

As of 2017, The Walney Wind Farm in the United Kingdom is the largest offshore wind farm in the world at 659 MW, followed by the London Array (630 MW) also in the UK.

Offshore wind turbines are less obtrusive than turbines on land, as their apparent size and noise is mitigated by distance. Because water has less surface roughness than land (especially deeper water), the average wind speed is usually considerably higher over open water. Capacity factors (utilisation rates) are considerably higher than for onshore locations.

The province of Ontario in Canada is pursuing several proposed locations in the Great Lakes, including the suspended Trillium Power Wind 1 approximately 20 km from shore and over 400 MW in size. Other Canadian projects include one on the Pacific west coast.

In 2010, there were no offshore wind farms in the United States, but projects were under development in wind-rich areas of the East Coast, Great Lakes, and Pacific coast; and in late 2016 the Block Island Wind Farm was commissioned.

Installation and service/maintenance of off-shore wind farms are a specific challenge for technology and economic operation of a wind farm. As of 2015, there are 20 jackup vessels for lifting components, but few can lift sizes above 5MW. Service vessels have to be operated nearly 24/7 (availability higher than 80% of time) to get sufficient amortisation from the wind turbines. Therefore, special fast service vehicles for installation (like Wind Turbine Shuttle) as well as for maintenance (including heave compensation and heave compensated working platforms to allow the service staff to enter the wind turbine also at difficult weather conditions) are required. So-called inertial and optical based Ship Stabilization and Motion Control systems (iSSMC) are used for that.

Experimental and Proposed Wind Farms

Experimental wind farms consisting of a single wind turbine for testing purposes have been built. One such installation is Østerild Wind Turbine Test Field.

Airborne wind farms have been envisaged. Such wind farms are a group of airborne wind energy systems located close to each other connected to the grid at the same point.

Wind farms consisting of diverse wind turbines have been proposed in order to efficiently use wider ranges of wind speeds. Such wind farms are proposed to be projected under two criteria: maximization of the energy produced by the farm and minimization of its costs.

Criticism

Environmental Impact

Livestock near a wind turbine.

Compared to the environmental impact of traditional energy sources, the environmental impact of wind power is relatively minor. Wind power consumes no fuel, and emits no air pollution, unlike fossil fuel power sources. The energy consumed to manufacture and transport the materials used to build a wind power plant is equal to the new energy produced by the plant within a few months.

Onshore wind farms are criticized for their impact on the landscape. Their network of turbines, roads, transmission lines and substations can result in "energy sprawl". Typically they need to take up more land than other power stations and are more spread out. To power many major cities by wind alone would require building wind farms bigger than the cities themselves. Typically they also need to be built in wild and rural areas, which can lead to "industrialization of the countryside" and habitat loss. A report by the Mountaineering Council of Scotland concluded that wind farms have a negative impact on tourism in areas known for natural landscapes and panoramic views. However, land between the turbines can still be used for agriculture.

Habitat loss and habitat fragmentation are the greatest impact of wind farms on wildlife. There are also reports of higher bird and bat mortality at wind turbines as there are around other artificial structures. The scale of the ecological impact may or may not be significant, depending on specific circumstances. The estimated number of bird deaths caused by wind turbines in the United States is between 140,000 and 328,000, whereas deaths caused by domestic cats in the United States are estimated to be between 1.3 and 4.0 billion birds each year and over 100 million birds are killed in the United States each year by impact with windows. Prevention and mitigation of wildlife fatalities, and protection of peat bogs, affect the siting and operation of wind turbines.

Human Health

Wind turbines overlooking Ardrossan, Scotland.

There have been multiple scientific, peer-reviewed studies into wind farm noise, which have concluded that infrasound from wind farms is not a hazard to human health and there is no verifiable evidence for 'Wind Turbine Syndrome' causing Vibroacoustic disease, although some suggest further research might still be useful.

A 2007 report by the U.S. National Research Council noted that noise produced by wind turbines is generally not a major concern for humans beyond a half-mile or so. Low-frequency vibration and its effects on humans are not well understood and sensitivity to such vibration resulting from wind-turbine noise is highly variable among humans. There are opposing views on this subject, and more research needs to be done on the effects of low-frequency noise on humans.

In a 2009 report about "Rural Wind Farms", a Standing Committee of the Parliament of New South Wales, Australia, recommended a minimum setback of two kilometres between wind turbines and neighbouring houses (which can be waived by the affected neighbour) as a precautionary approach.

A 2014 paper suggests that the 'Wind Turbine Syndrome' is mainly caused by the nocebo effect and other psychological mechanisms. Australian science magazine Cosmos states that although the symptoms are real for those who suffer from the condition, doctors need to first eliminate known causes (such as pre-existing cancers or thyroid disease) before reaching definitive conclusions with the caveat that new technologies often bring new, previously unknown health risks.

Effect on Power Grid

Utility-scale wind farms must have access to transmission lines to transport energy. The wind farm developer may be obliged to install extra equipment or control systems in the wind farm to meet the technical standards set by the operator of a transmission line.

The intermittent nature of wind power can pose complications for maintaining a stable power grid when wind farms provide a large percentage of electricity in any one region.

Ground Radar Interference

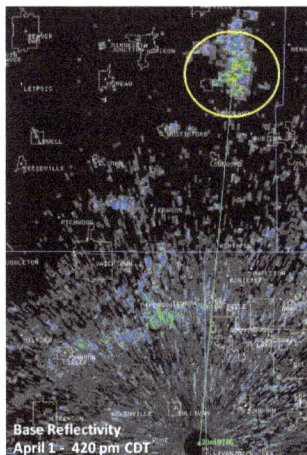

Wind farm interference (in yellow circle) on radar map.

Wind farms can interfere with ground radar systems used for military, weather and air traffic control. The large, rapidly moving blades of the turbines can return signals to the radar that can be mistaken as an aircraft or weather pattern. Actual aircraft and weather patterns around wind farms can be accurately detected, as there is no fundamental physical constraint preventing that. But aging radar infrastructure is significantly challenged with the task. The US military is using wind turbines on some bases, including Barstow near the radar test facility.

Effects

The level of interference is a function of the signal processors used within the radar, the speed of the aircraft and the relative orientation of wind turbines/aircraft with respect to the radar. An aircraft flying above the wind farm's turning blades could become impossible to detect because the blade tips can be moving at nearly aircraft velocity. Studies are currently being performed to determine the level of this interference and will be used in future site planning. Issues include masking (shadowing), clutter (noise), and signal alteration. Radar issues have stalled as much as 10,000 MW of projects in USA.

Some very long range radars are not affected by wind farms.

Mitigation

Permanent problem solving include a *non-initiation window* to hide the turbines while still tracking aircraft over the wind farm, and a similar method mitigates the false returns. England's Newcastle Airport is using a short-term mitigation; to "blank" the turbines on the radar map with a software patch. Wind turbine blades using stealth technology are being developed to mitigate radar reflection problems for aviation. As well as stealth windfarms, the future development of infill radar systems could filter out the turbine interference.

A mobile radar system, the Lockheed Martin TPS-77, can distinguish between aircraft and wind turbines, and more than 170 TPS-77 radars are in use around the world.

Radio Reception Interference

There are also reports of negative effects on radio and television reception in wind farm communities. Potential solutions include predictive interference modelling as a component of site selection.

Wind turbines can often cause terrestrial television interference when the direct path between television transmitter and receiver is blocked by terrain. Interference effects become significant when the reflected signal from the turbine blades approaches the strength of the direct unreflected signal. Reflected signals from the turbine blades can cause loss of picture, pixellation and disrupted sound. There is a common misunderstanding that digital TV signals will not be affected by turbines — in practice they are.

Agriculture

A 2010 study found that in the immediate vicinity of wind farms, the climate is cooler during the day and slightly warmer during the night than the surrounding areas due to the turbulence generated by the blades.

In another study an analysis carried out on corn and soybean crops in the central areas of the United States noted that the microclimate generated by wind turbines improves crops as it prevents the late spring and early autumn frosts, and also reduces the action of pathogenic fungi that grow on the leaves. Even at the height of summer heat, the lowering of 2.5–3 degrees above the crops due to turbulence caused by the blades, can make a difference for the cultivation of corn.

OFFSHORE WIND POWER

Wind turbines and electrical substation of Alpha Ventus Offshore Wind Farm in the North Sea.

Offshore wind power or offshore wind energy is the use of wind farms constructed in bodies of water, usually in the ocean on the continental shelf, to harvest wind energy to generate electricity. Higher wind speeds are available offshore compared to on land, so offshore wind power's electricity generation is higher per amount of capacity installed, and NIMBY opposition to construction is usually much weaker. Unlike the typical use of the term "offshore" in the marine industry, offshore wind power includes inshore water areas such as lakes, fjords and sheltered coastal areas, utilizing traditional fixed-bottom wind turbine technologies, as well as deeper-water areas utilizing floating wind turbines.

At the end of 2017, the total worldwide offshore wind power capacity was 18.8 GW. All the largest offshore wind farms are currently in northern Europe, especially in the United Kingdom and Germany, which together account for over two thirds of the total offshore wind power installed worldwide. As of September 2018, the 659 MW Walney Extension in the United Kingdom is the largest offshore wind farm in the world. The Hornsea Wind Farm under construction in the United Kingdom will become the largest when completed, at 1,200 MW. Other projects are in the planning stage, including Dogger Bank in the United Kingdom at 4.8 GW, and Greater Changhua in Taiwan at 2.4 GW.

The cost of offshore wind power has historically been higher than that of onshore wind generation, but costs have been decreasing rapidly in recent years and in Europe has been price-competitive with conventional power sources since 2017.

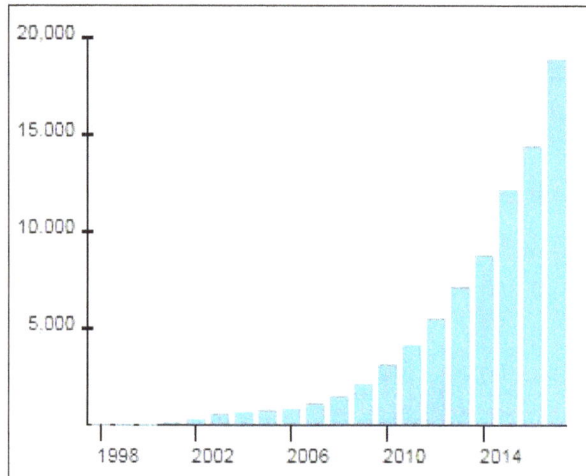
Global cumulative offshore capacity (MW).

An illustration of a hypothetical offshore wind farm in 1977.

Europe is the world leader in offshore wind power, with the first offshore wind farm (Vindeby) being installed in Denmark in 1991. In 2009, the average nameplate capacity of an offshore wind turbine in Europe was about 3 MW, and the capacity of future turbines was expected to increase to 5 MW.

In 2010, the US Energy Information Agency said "offshore wind power is the most expensive energy generating technology being considered for large scale deployment". The 2010 state of offshore wind power presented economic challenges significantly greater than onshore systems, with prices in the range of 2.5-3.0 million Euro/MW. That year, Siemens and Vestas were turbine suppliers for 90% of offshore wind power, while DONG Energy, Vattenfall and E.on were the leading offshore operators.

In 2011, DONG Energy estimated that while offshore wind turbines were not yet competitive with fossil fuels, they would be in 15 years. Until then, state funding and pension funds would be needed. At the end of 2011, there were 53 European offshore wind farms in waters off Belgium, Denmark, Finland, Germany, Ireland, the Netherlands, Norway, Sweden and the United Kingdom, with an operating capacity of 3,813 MW, while 5,603 MW was under construction. Offshore wind farms worth €8.5 billion ($11.4 billion) were under construction in European waters in 2011.

In 2012, Bloomberg estimated that energy from offshore wind turbines cost €161 (US$208) per MWh.

A 2013 comprehensive review of the engineering aspects of turbines like the sizes used onshore, including the electrical connections and converters, considered that the industry had in general been overoptimistic about the benefits-to-costs ratio and concluded that the "offshore wind market doesn't look as if it is going to be big". In 2013, offshore wind power contributed to 1,567 MW of the total 11,159 MW of wind power capacity constructed that year.

By January 2014, 69 offshore wind farms had been constructed in Europe with an average annual rated capacity of 482 MW. The total installed capacity of offshore wind farms in European waters reached 6,562 MW. The United Kingdom had by far the largest capacity with 3,681 MW. Denmark was second with 1,271 MW installed and Belgium was third with 571 MW. Germany came fourth with 520 MW, followed by the Netherlands (247 MW), Sweden (212 MW), Finland (26 MW), Ireland (25 MW), Spain (5 MW), Norway (2 MW) and Portugal (2 MW).

At the end of 2015, 3,230 turbines at 84 offshore wind farms across 11 European countries had been installed and grid-connected, making a total capacity of 11,027 MW.

Outside of Europe, the Chinese government had set ambitious targets of 5 gigawatt (GW) of installed offshore wind capacity by 2015 and 30 GW by 2020 that would eclipse capacity in other countries. However, in May 2014 the capacity of offshore wind power in China was only 565 MW. Offshore capacity in China increased by 832 MW in 2016, of which 636 MW were made in China.

The offshore wind construction market remains quite concentrated. By the end of 2015, Siemens Wind Power had installed 63% of the world's 11 GW offshore wind power capacity; Vestas had 19%, Senvion came third with 8% and Adwen 6%. About 12 GW of offshore wind power capacity was operational, mainly in Northern Europe, with 3,755 MW of that coming online during 2015.

Costs of offshore wind power are decreasing much faster than expected. By 2016, four contracts (Borssele and Kriegers) were already below the lowest of the predicted 2050 prices.

Projections for 2020 estimate an offshore wind farm capacity of 40 GW in European waters, which would provide 4% of the European Union's demand of electricity. The European Wind Energy Association has set a target of 40 GW installed by 2020 and 150 GW by 2030. Offshore wind power capacity is expected to reach a total of 75 GW worldwide by 2020, with significant contributions from China and the United States.

The Organisation for Economic Co-operation and Development (OECD) predicted in 2016 that offshore wind power will grow to 8% of ocean economy by 2030, and that its industry will employ 435,000 people, adding $230 billion of value.

Types of Offshore Wind Turbines

Fixed Foundation Offshore Wind Turbines

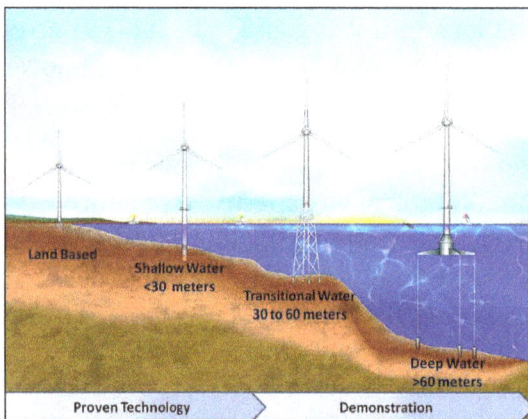

Progression of expected wind turbine evolution to deeper water.

Tripods foundation for offshore wind farms in 2008 in Wilhelmshaven, Germany.

Almost all currently operating offshore wind farms employ fixed foundation turbines, with the exception of a few pilot projects. Fixed foundation offshore wind turbines have fixed foundations underwater, and are installed in relatively shallow waters of up to 50–60 m.

Types of underwater structures include monopile, tripod, and jacketed, with various foundations at the sea floor including monopile or multiple piles, gravity base, and caissons. Offshore turbines require different types of bases for stability, according to the depth of water. To date a number of different solutions exist:

- A monopile (single column) base, six meters in diameter, is used in waters up to 30 meters deep.

- Gravity base structures, for use at exposed sites in water 20–80 m deep.

- Tripod piled structures, in water 20–80 m deep.

- Tripod suction caisson structures, in water 20–80 m deep.

- Conventional steel jacket structures, as used in the oil and gas industry, in water 20–80 m deep.

Monopiles up to 11 m diameter at 2,000 tonnes can be made, but the largest so far are 1,300 tons which is below the 1,500 tonnes limit of some crane vessels. The other turbine components are much smaller.

The tripod pile substructure system is a new concept developed to reach deeper waters than with the shallow water systems, up to 60 m. This technology consists of three monopiles linked together through a joint piece at the top. The main advantage of this solution is the simplicity of the installation, which is done by installing three monopiles and then adding the upper joint.

Tripod is an innovative concept that consists on a central pipe that lies on a tripod tubular frame configuration at its bottom part. This uses three small seabed driven piles at each leg of the tripod to link it to the seabed. The main advantage of the tripod system is that it has a larger base, which decreases its risk of getting overturned. Due to the large dimensions the installation process is more difficult and increases the cost.

A steel jacket structure comes from an adaptation to the offshore wind industry of concepts that have been in use in the oil and gas industry for decades. Their main advantage lies in the possibility of reaching higher depths (up to 80m). Their main limitations are due to the high construction and installation costs.

Floating Offshore Wind Turbines

For locations with depths over about 60–80 m, fixed foundations are uneconomical or technically unfeasible, and floating wind turbine anchored to the ocean floor are needed. *Hywind* is the world's first full-scale floating wind turbine, installed in the North Sea off Norway in 2009. Hywind Scotland, commissioned in October 2017, is the first operational floating wind farm, with a capacity of 30 MW. Other kinds of floating turbines have been deployed, and more projects are planned.

Vertical Axis Offshore Wind Turbines

Although the great majority of onshore and all large-scale offshore wind turbines currently installed are horizontal axis, vertical axis wind turbines have been proposed for use in offshore installations. Thanks to the installation offshore and their lower center of gravity, these turbines can in principle be built bigger than horizontal axis turbines, with proposed designs of up to 20 MW capacity per turbine. This could improve the economy of scale of offshore wind farms. However, there are no current large-scale demonstrations of this technology.

Economics

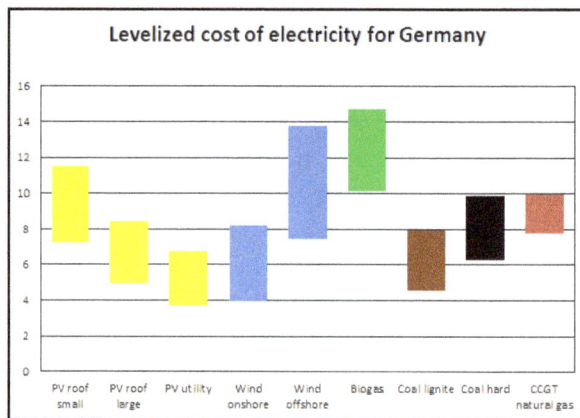

Comparison of the levelized cost of electricity of offshore wind power compared to other sources in Germany in 2018.

The advantage of locating wind turbines offshore is that the wind is much stronger off the coasts, and unlike wind over the continent, offshore breezes can be strong in the afternoon, matching the time when people are using the most electricity. Offshore turbines can also be located close to the load centers along the coasts, such as large cities, eliminating the need for new long-distance transmission lines. However, there are several disadvantages of offshore installations, related to more expensive installation, difficulty of access, and harsher conditions for the units.

Locating wind turbines offshore exposes the units to high humidity, salt water and salt water spray which negatively affect service life, cause corrosion and oxidation, increase maintenance and repair costs and in general make every aspect of installation and operation much more difficult, time-consuming, more dangerous and far more expensive than sites on land. The humidity and temperature is controlled by air conditioning the sealed nacelle. Sustained high-speed operation and generation also increases wear, maintenance and repair requirements proportionally.

The cost of the turbine represents just one third to one half of total costs in offshore projects today, the rest comes from infrastructure, maintenance, and oversight. Costs for foundations, installation, electrical connections and operation and maintenance

(O&M) are a large share of the total for offshore installations compared to onshore wind farms. The cost of installation and electrical connection also increases rapidly with distance from shore and water depth.

Other limitations of offshore wind power are related to the still limited number of installations. The offshore wind industry is not yet fully industrialized, as supply bottlenecks still exist as of 2017.

Investment Costs

Offshore wind farms tend to have larger turbines when compared to onshore installations, and the trend is towards a continued increase in size. Economics of offshore wind farms tend to favor larger turbines, as installation and grid connection costs decrease per unit energy produced. Moreover, offshore wind farms do not have the same restriction in size of onshore wind turbines, such as availability of land or transportation requirements.

Operating Costs

Operational expenditures for wind farms are split between Maintenance (38%), Port Activities (31%), Operation (15%), License Fees (12%), and Miscellaneous Costs (4%).

Operation and maintenance costs typically represent 53% of operational expenditures, and 25% - 30% of the total lifecycle costs for offshore wind farms. O&Ms are considered one of the major barriers for further development of this resource.

Maintenance of offshore wind farms is much more expensive than for onshore installations. For example, a single technician in a pickup truck can quickly, easily and safely access turbines on land in almost any weather conditions, exit his or her vehicle and simply walk over to and into the turbine tower to gain access to the entire unit within minutes of arriving onsite. Similar access to offshore turbines involves driving to a dock or pier, loading necessary tools and supplies into boat, a voyage to the wind turbine(s), securing the boat to the turbine structure, transferring tools and supplies to and from boat to turbine and turbine to boat and performing the rest of the steps in reverse order. In addition to standard safety gear such as a hardhat, gloves and safety glasses, an offshore turbine technician may be required to wear a life vest, waterproof or water-resistant clothing and perhaps even a survival suit if working, sea and atmospheric conditions make rapid rescue in case of a fall into the water unlikely or impossible. Typically at least two technicians skilled and trained in operating and handling large power boats at sea are required for tasks that one technician with a driver's license can perform on land in a fraction of the time at a fraction of the cost.

Cost of Energy

Auctions in 2016 have reached costs of €54.5 per MWh at the 700 MW Borssele 3&4 due to government tender and size, and €49.90 per MWh (without transmission) at the 600 MW Kriegers Flak.

In September 2017 contracts were awarded in the UK for a strike price of £57.50 per MWh making the price cheaper than nuclear and competitive with gas.

In September 2018 contracts were awarded for Vineyard Wind, Massachusetts, USA at a cost of between $65-$74 per MWh.

Offshore Wind Resources

Offshore wind resource characteristics span a range of spatial and temporal scales and field data on external conditions. For the North Sea, wind turbine energy is around 30 kWh/m^2 of sea area, per year, delivered to grid. The energy per sea area is roughly independent of turbine size.

Planning and Permitting

A number of things are necessary in order to attain the necessary information for planning the commissioning of an offshore wind farm. The first information required is offshore wind characteristics. Additional necessary data for planning includes water depth, currents, seabed, migration, and wave action, all of which drive mechanical and structural loading on potential turbine configurations. Other factors include marine growth, salinity, icing, and the geotechnical characteristics of the sea or lake bed.

Existing hardware for measurements includes Light Detection and Ranging (LIDAR), Sonic Detection and Ranging (SODAR), radar, autonomous underwater vehicles (AUV), and remote satellite sensing, although these technologies should be assessed and refined, according to a report from a coalition of researchers from universities, industry, and government, supported by the Atkinson Center for a Sustainable Future.

Because of the many factors involved, one of the biggest difficulties with offshore wind farms is the ability to predict loads. Analysis must account for the dynamic coupling between translational (surge, sway, and heave) and rotational (roll, pitch, and yaw) platform motions and turbine motions, as well as the dynamic characterization of mooring lines for floating systems. Foundations and substructures make up a large fraction of offshore wind systems, and must take into account every single one of these factors. Load transfer in the grout between tower and foundation may stress the grout, and elastomeric bearings are used in several British sea turbines.

Corrosion is also a serious problem and requires detailed design considerations. The prospect of remote monitoring of corrosion looks very promising using expertise utilised by the offshore oil/gas industry and other large industrial plants.

Some of the guidelines for designing offshore wind farms are IEC 61400-3, but in the US several other standards are necessary. In the EU, different national standards are to be straightlined into more cohesive guidelines to lower costs. The standards requires that a loads analysis is based on site-specific external conditions such as wind, wave and currents.

The planning and permitting phase can cost more than $10 million, take 5–7 years and have an uncertain outcome. The industry puts pressure on the governments to improve the processes. In Denmark, many of these phases have been deliberately streamlined by authorities in order to minimize hurdles, and this policy has been extended for coastal wind farms with a concept called 'one-stop-shop'. The United States introduced a similar model called "Smart from the Start" in 2012.

Installation

Several foundation structures for offshore wind turbines in a port.

Specialized jackup rigs (Turbine Installation Vessels) are used to install foundation and turbine. As of 2019 the next generation of vessels are being built, capable of lifting 3-5,000 tons to 160 meters.

A large number of monopile foundations have been utilized in recent years for economically constructing fixed-bottom offshore wind farms in shallow-water locations. Each utilizes a single, generally large-diameter, foundation structural element to support all the loads (weight, wind, etc.) of a large above-surface structure.

The typical construction process for a wind turbine sub-sea monopile foundation in sand includes using a pile driver to drive a large hollow steel pile 25 m deep into the seabed, through a 0.5 m layer of larger stone and gravel to minimize erosion around the pile. These piles can be 4 m in diameter with approximately 50mm thick walls. A transition piece (complete with pre-installed features such as boat-landing arrangement, cathodic protection, cable ducts for sub-marine cables, turbine tower flange, etc.) is attached to the now deeply driven pile, the sand and water are removed from the centre of the pile and replaced with concrete. An additional layer of even larger stone, up to 0.5 m diameter, is applied to the surface of the seabed for longer-term erosion protection.

Grid Integration

There are several different types of technologies that are being explored as viable options for integrating offshore wind power into the onshore grid. The most conventional method is through high-voltage alternating current (HVAC) transmission lines. HVAC transmission lines are currently the most commonly used form of grid connections for offshore wind turbines. However, there are significant limitations that prevent HVAC from being practical, especially as the distance to offshore turbines increases. First, HVAC is limited by cable charging currents, which are a result of capacitance in the cables. Undersea AC cables have a much higher capacitance than overhead AC cables, so losses due to capacitance become much more significant, and the voltage magnitude at the receiving end of the transmission line can be significantly different from the magnitude at the receiving end. In order to compensate for these losses, either more cables or reactive compensation must be added to the system. Both of these add costs to the system. Additionally, because HVAC cables have both real and reactive power flowing through them, there can be additional losses. Because of these losses, underground HVAC lines are limited in how far they can extend. The maximum appropriate distance for HVAC transmission for offshore wind power is considered to be around 80 km.

An offshore structure for housing an HVDC converter station for offshore
wind parks is being moved by a heavy-lift ship in Norway.

Using high-voltage direct current (HVDC) cables has been a proposed alternative to using HVAC cables. HVDC transmission cables are not affected by the cable charging currents and experience less power loss because HVDC does not transmit reactive power. With less losses, undersea HVDC lines can extend much farther than HVAC. This makes HVDC preferable for siting wind turbines very far offshore. However, HVDC requires power converters in order to connect to the AC grid. Both line commutated converters (LCCs) and voltage source converters (VSCs) have been considered for this. Although LCCs are a much more widespread technology and cheaper, VSCs have many more benefits, including independent active power and reactive power control. New research has been put into developing hybrid HVDC technologies that have a LCC connected to a VSC through a DC cable.

Maintenance

Offshore wind turbines of the Rødsand Wind Farm in the Fehmarn Belt,
the western part of the Baltic Sea between Germany and Denmark.

Turbines are much less accessible when offshore (requiring the use of a service vessel or helicopter for routine access, and a jackup rig for heavy service such as gearbox replacement), and thus reliability is more important than for an onshore turbine. Some wind farms located far from possible onshore bases have service teams living on site in offshore accommodation units. To limit the effects of corrosion on the blades of a wind turbine, a protective tape of elastomeric materials is applied, though the droplet erosion protection coatings provide better protection from the elements.

A maintenance organization performs maintenance and repairs of the components, spending almost all its resources on the turbines. The conventional way of inspecting the blades is for workers to rappel down the blade, taking a day per turbine. Some farms inspect the blades of three turbines per day by photographing them from the monopile through a 600mm lens, avoiding to go up. Others use camera drones.

Because of their remote nature, prognosis and health-monitoring systems on offshore wind turbines will become much more necessary. They would enable better planning just-in-time maintenance, thereby reducing the operations and maintenance costs. According to a report from a coalition of researchers from universities, industry, and government (supported by the Atkinson Center for a Sustainable Future), making field data from these turbines available would be invaluable in validating complex analysis codes used for turbine design. Reducing this barrier would contribute to the education of engineers specializing in wind energy.

Decommissioning

As the first offshore wind farms reach their end of life, a demolition industry develops to recycle them at a cost of DKK 2-4 million per MW, to be guaranteed by the owner. The first offshore wind farm to be decommissioned was Yttre Stengrund in Sweden in November 2015, followed by Vindeby in 2017 and Blyth in 2019.

Advantages and Disadvantages of Offshore Wind Farms

Advantages

- Offshore wind speeds tend to be faster than on land.1 Small increases in wind speed yield large increases in energy production: a turbine in a 15-mph wind can generate twice as much energy as a turbine in a 12-mph wind. Faster wind speeds offshore mean much more energy can be generated.

- Offshore wind speeds tend to be steadier than on land. A steadier supply of wind means a more reliable source of energy.

- Many coastal areas have very high energy needs. Half of the United States' population lives in coastal areas, with concentrations in major coastal cities. Building offshore wind farms in these areas can help to meet those energy needs from nearby sources.

- Offshore wind farms have many of the same advantages as land-based wind farms – they provide renewable energy; they do not consume water; they provide a domestic energy source; they create jobs; and they do not emit environmental pollutants or greenhouse gases.

Disadvantages

- Offshore wind farms can be expensive and difficult to build and maintain. In particular:

  ◦ It is very hard to build robust and secure wind farms in water deeper than around 200 feet (~60 m), or over half a football field's length. Although coastal waters off the east coast of the U.S. are relatively shallow, almost all of the potential wind energy resources off the west coast are in waters exceeding this depth. Floating wind turbines are beginning to overcome this challenge.

  ◦ Wave action, and even very high winds, particularly during heavy storms or hurricanes, can damage wind turbines.

  ◦ The production and installation of power cables under the seafloor to transmit electricity back to land can be very expensive.

- Effects of offshore wind farms on marine animals and birds are not fully understood.

- Offshore wind farms built within view of the coastline (up to 26 miles offshore, depending on viewing conditions) may be unpopular among local residents, and may affect tourism and property values.

References

- Advantages-and-disadvantages-of-wind-energy, wind-energy, wind: clean-energy-ideas.com, Retrieved 16 January, 2019

- Egration of offshore wind farms – case studies. Wiley. Doi:10.1002/9781118701638.ch5. Isbn 9781118701638

- Windturbines: conserve-energy-future.com, Retrieved 29 March, 2019

- Watts, jonathan & huang, cecily. Winds of change blow through china as spending on renewable energy soars, the guardian, 19 march 2012, revised on 20 march 2012. Retrieved 4 january 2012.

- Begoña guezuraga; rudolf zauner; werner pölz (january 2012). "life cycle assessment of two different 2 mw class wind turbines". Renewable energy. 37 (1): 37. Doi:10.1016/j.renene.2011.05.008

- Offshore wind farms : technologies, design and operation. Ng, chong,, ran, li,. Duxford, uk. Isbn 978-0-08-100780-8. Oclc 944186047

4

Geothermal Energy and Technologies

The heat which comes from the sub-surface of the Earth is known as geothermal energy. Various geothermal technologies are used to harness the Earth's heat. The main types of such technologies are ground source heat pumps, direct use geothermal and deep and enhanced geothermal systems. These applications of geothermal energy have been thoroughly discussed in this chapter.

Geothermal energy is a form of energy conversion in which heat energy from within Earth is captured and harnessed for cooking, bathing, space heating, electrical power generation, and other uses.

Heat from Earth's interior generates surface phenomena such as lava flows, geysers, fumaroles, hot springs, and mud pots. The heat is produced mainly by the radioactive decay of potassium, thorium, and uranium in Earth's crust and mantle and also by friction generated along the margins of continental plates. The subsequent annual low-grade heat flow to the surface averages between 50 and 70 milliwatts (mW) per square metre worldwide. In contrast, incoming solar radiation striking Earth's surface provides 342 watts per square metre annually. Geothermal heat energy can be recovered and exploited for human use, and it is available anywhere on Earth's surface. The estimated energy that can be recovered and utilized on the surface is 4.5×10^6 exajoules, or about 1.4×10^6 terawatt-years, which equates to roughly three times the world's annual consumption of all types of energy.

The amount of usable energy from geothermal sources varies with depth and by extraction method. The increase in temperature of rocks and other materials underground averages 20–30 °C (36–54 °F) per kilometre (0.6 mile) depth worldwide in the upper part of the lithosphere, and this rate of increase is much higher in most of Earth's known geothermal areas. Normally, heat extraction requires a fluid (or steam) to bring the energy to the surface. Locating and developing geothermal resources can be challenging. This is especially true for the high-temperature resources needed for generating electricity. Such resources are typically limited to parts of the world characterized by recent volcanic activity or located along plate boundaries or within crustal hot spots. Even though there is a continuous source of heat within Earth, the extraction rate of the heated fluids and steam can exceed the replenishment rate, and, thus, use of the resource must be managed sustainably.

Uses

Geothermal energy use can be divided into three categories: direct-use applications, geothermal heat pumps (GHPs), and electric power generation.

Direct Uses

Probably the most widely used set of applications involves the direct use of heated water from the ground without the need for any specialized equipment. All direct-use applications make use of low-temperature geothermal resources, which range between about 50 and 150 °C (122 and 302 °F). Such low-temperature geothermal water and steam have been used to warm single buildings, as well as whole districts where numerous buildings are heated from a central supply source. In addition, many swimming pools, balneological (therapeutic) facilities at spas, greenhouses, and aquaculture ponds around the world have been heated with geothermal resources. Other direct uses of geothermal energy include cooking, industrial applications (such as drying fruit, vegetables, and timber), milk pasteurization, and large-scale snow melting. For many of those activities, hot water is often used directly in the heating system, or it may be used in conjunction with a heat exchanger, which transfers heat when there are problematic minerals and gases such as hydrogen sulfide mixed in with the fluid.

Bagno Vignoni: hot springs
Hot springs in Bagno Vignoni, Italy.

Geothermal Heat Pumps

Geothermal heat pumps (GHPs) take advantage of the relatively stable moderate temperature conditions that occur within the first 300 metres (1,000 feet) of the surface to heat buildings in the winter and cool them in the summer. In that part of the lithosphere, rocks and groundwater occur at temperatures between 5 and 30 °C (41 and 86 °F). At shallower depths, where most GHPs are found, such as within 6 metres (about 20 feet) of Earth's surface, the temperature of the ground maintains a near-constant temperature of 10 to 16 °C (50 to 60 °F). Consequently, that heat can be used to help warm buildings during the colder months of the year when the air temperature falls below that of the ground. Similarly, during the warmer months of the year, warm air can be drawn from a building and circulated underground, where it loses much of its heat and is returned.

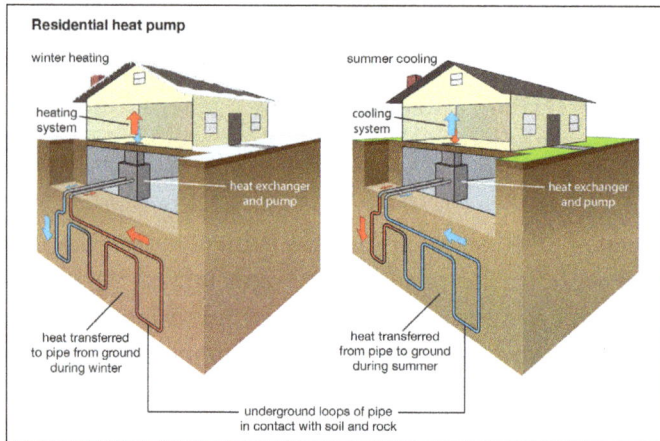

Residential heat pump:
Residential heat pump operation for summer cooling and winter heating.

A GHP system is made up of a heat exchanger (a loop of pipes buried in the ground) and a pump. The heat exchanger transfers heat energy between the ground and air at the surface by means of a fluid that circulates through the pipes; the fluid used is often water or a combination of water and antifreeze. During warmer months, heat from warm air is transferred to the heat exchanger and into the fluid. As it moves through the pipes, the heat is dispersed to the rocks, soil, and groundwater. The pump is reversed during the colder months. Heat energy stored in the relatively warm ground raises the temperature of the fluid. The fluid then transfers this energy to the heat pump, which warms the air inside the building.

GHPs have several advantages over more conventional heating and air-conditioning systems. They are very efficient, using 25–50 percent less electricity than comparable conventional heating and cooling systems, and they produce less pollution. The reduction in energy use associated with GHPs can translate into as much as a 44 percent decrease in greenhouse gas emissions compared with air-source heat pumps (which transfer heat between indoor and outdoor air). In addition, when compared with electric resistance heating systems (which convert electricity to heat) coupled with standard air-conditioning systems, GHPs can produce up to 72 percent less greenhouse gas emissions.

Electric Power Generation

Depending upon the temperature and the fluid (steam) flow, geothermal energy can be used to generate electricity. Geothermal power plants can produce electricity in three ways. Despite their differences in design, all three control the behaviour of steam and use it to drive electrical generators. Given that the excess water vapour at the end of each process is condensed and returned to the ground, where it is reheated for later use, geothermal power is considered a form of renewable energy.

Some geothermal power plants simply collect rising steam from the ground. In such "dry steam" operations, the heated water vapour is funneled directly into a turbine that drives an electrical generator. Other power plants, built around the flash steam and binary cycle designs, use a mixture of steam and heated water ("wet steam") extracted from the ground to start the electrical generation process.

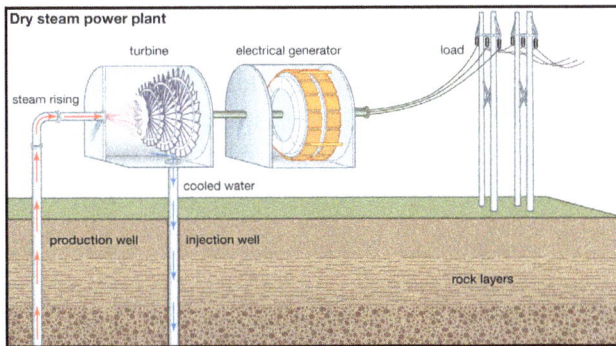

Dry steam geothermal power generation.

In flash steam power plants, pressurized high-temperature water is drawn from beneath the surface into containers at the surface, called flash tanks, where the sudden decrease in pressure causes the liquid water to "flash," or vaporize, into steam. The steam is then used to power the turbine-generator set. In contrast, binary-cycle power plants use steam driven off a secondary working fluid (such as ammonia and hydrocarbons) contained within a closed loop of pipes to power the turbine-generator set. In this process, geothermally heated water is drawn up through a different set of pipes, and much of the energy stored in the heated water is transferred to the working fluid through a heat exchanger. The working fluid then vaporizes. After the vapour from the working fluid passes through the turbine, it is recondensed and piped back to the heat exchanger.

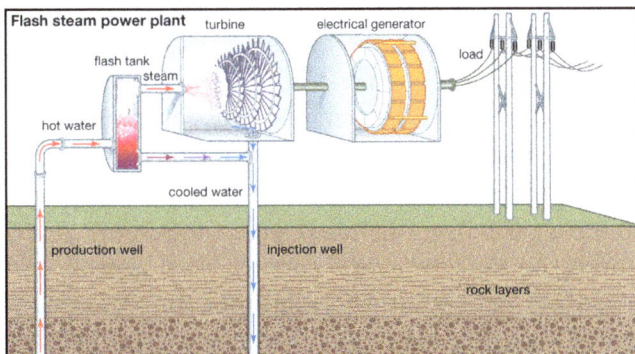

Flash steam geothermal power generation.

Electrical power usually requires water heated above 175 °C (347 °F) to be economical. In geothermal plants using the Organic Rankine Cycle (ORC), a special type of binary-cycle technology that utilizes lower-temperature heat sources (such as biomass combustion and industrial waste heat), water temperatures as low as 85–90 °C (185–194 °F) may be used.

Extraction

Geothermal energy is best found in areas with high thermal gradients. Those gradients occur in regions affected by recent volcanism, in areas located along plate boundaries (such as along the Pacific Ring of Fire), or in areas marked by thin crust (hot spots) such as Yellowstone National Park and the Hawaiian Islands. Geothermal reservoirs associated with those regions must have a heat source, adequate water recharge, a reservoir with adequate permeability or faults that allow fluids to rise close to the surface, and an impermeable caprock to prevent the escape of the heat. In addition, such reservoirs must be economically accessible (that is, within the range of drills).

A geothermal power station in Iceland that creates electricity
from heat generated in Earth's interior.

The heated fluid from a geothermal resource is tapped by drilling wells, sometimes as deep as 9,100 metres (about 30,000 feet), and is extracted by pumping or by natural artesian flow (where the weight of the water forces it to the surface). Water and steam are then piped to the power plant to generate electricity or through insulated pipelines—which may be buried or placed aboveground—for use in heating and cooling applications. In general, electric power plant pipelines are limited to roughly 1.6 km (1 mile) in length to minimize heat loss in the steam. However, direct-use pipelines spanning several tens of kilometres have been installed with a temperature loss of less than 2–5 °C (3.6–9 °F), depending on the flow rate. The most economically efficient facilities are located close to the geothermal resource to minimize the expense of constructing long pipelines. In the case of electric power generation, costs can be kept down by locating the facility near electrical transmission lines to transmit the electricity to market.

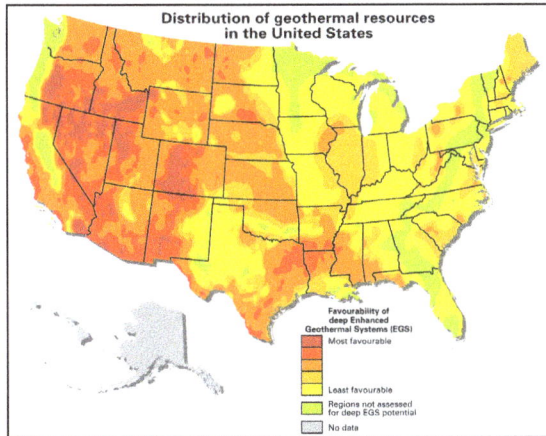

Geothermal energy potential
Map of geothermal energy resources in the United States.

Exhaustion

Geothermal resources can be exhausted if the rate of heat extraction exceeds the rate of natural heat recharge. Normally, geothermal resources can be used for 20 to 30 years; however, the energy output may decrease with time, making continued development uneconomical. On the other hand, geothermal electric power has been produced continually from the Larderello geothermal field since the early 1900s and at the Geysers since 1960. there has been a decline in both of those fields, this problem has been partially overcome by drilling new wells and by recharging the water supply. At the Geysers, electrical capacity declined from 1,800 MW to approximately 1,000 MW, but about 200 MW of capacity was returned by placing the field under one operator and constructing pipelines to deliver wastewater for recharging the reservoir. Projects such as the Reykjavík district heating system have been operating since the 1930s with little change in the output, and the Oregon Institute of Technology geothermal heating system has been operating since the 1950s with no change in production. Thus, with proper management, geothermal resources can be sustainable for many years, and they can even recover if use is suspended for a period of time.

Environmental Effects and Economic Costs

The environmental effects of geothermal development and power generation include the changes in land use associated with exploration and plant construction, noise and sight pollution, the discharge of water and gases, the production of foul odours, and soil subsidence. Most of those effects, however, can be mitigated with current technology so that geothermal uses have no more than a minimal impact on the environment. For example, Klamath Falls, Oregon, has approximately 600 geothermal wells for residential space heating. The city has also invested in a district heating system and a downtown snow-melting system, and it provides heating to local businesses. However, none of the systems used to supply and deliver geothermal energy are visible in town.

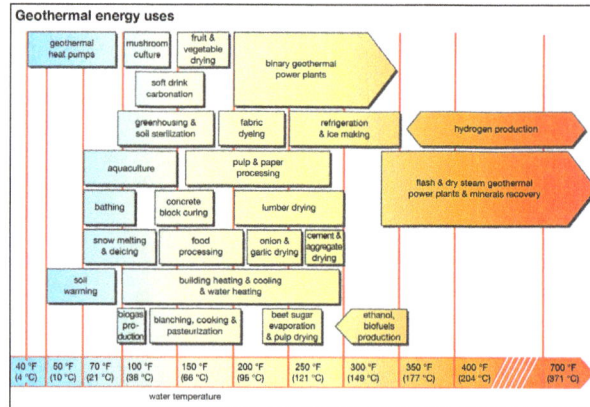

Geothermal energy uses

geothermal heat pumps | mushroom culture | fruit & vegetable drying

binary geothermal power plants

soft drink carbonation

greenhousing & soil sterilization | fabric dyeing | refrigeration & ice making

hydrogen production

aquaculture | pulp & paper processing

flash & dry steam geothermal power plants & minerals recovery

bathing | concrete block curing | lumber drying

snow melting & deicing | food processing | onion & garlic drying | cement & aggregate drying

soil warming | building heating & cooling & water heating

biogas production | blanching, cooking & pasteurization | beet sugar evaporation & pulp drying | ethanol, biofuels production

| 40 °F (4 °C) | 50 °F (10 °C) | 70 °F (21 °C) | 100 °F (38 °C) | 150 °F (66 °C) | 200 °F (95 °C) | 250 °F (121 °C) | 300 °F (149 °C) | 350 °F (177 °C) | 400 °F (204 °C) | 700 °F (371 °C) |

water temperature

Geothermal energy uses:
Diagram of various geothermal energy uses displayed according to the
water temperature of the geothermal resource.

In addition, GHPs have a very minimal effect on the environment, because they make use of shallow geothermal resources within 100 metres (about 330 feet) of the surface. GHPs cause only small temperature changes to the groundwater or rocks and soil in the ground. In closed-loop systems the ground temperature around the vertical boreholes is slightly increased or decreased; the direction of the temperature change is governed by whether the system is dominated by heating (which would be the case in colder regions) or cooling (which would be the case in warmer regions). With balanced heating and cooling loads, the ground temperatures will remain stable. Likewise, open-loop systems using groundwater or lake water would have very little effect on temperature, especially in regions characterized by high groundwater flows.

Comparing the benefits of geothermal energy with other renewable energy sources, the main advantage of geothermal energy is that its base load is available 24 hours per day, 7 days per week, whereas solar and wind are available only about one-third of the time. In addition, the cost of geothermal energy varies between 5 and 10 cents per kilowatt-hour, which can be competitive with other energy sources, such as coal. The main disadvantage of geothermal energy development is the high initial investment cost in constructing the facilities and infrastructure and the high risk of proving the resources. (Geothermal resources in low-permeability rocks are often found, and exploration activities often drill "dry" holes—that is, holes that produce steam in amounts too low to be exploited economically.) However, once the resource is proven, the annual cost of fuel (that is, hot water and steam) is low and tends not to escalate in price.

WORKING OF GEOTHERMAL ENERGY

Many regions of the world are already tapping geothermal energy as an affordable and sustainable solution to reducing dependence on fossil fuels, and the global warming and public health risks that result from their use. For example, as of 2013 more than 11,700 megawatts (MW) of large, utility-scale geothermal capacity was in operation globally, with another 11,700 MW in planned capacity additions on the way. These geothermal facilities produced approximately 68 billion kilowatt-hours of electricity, enough to meet the annual needs of more than 6 million typical U.S. households. Geothermal plants account for more than 25 percent of the electricity produced in both Iceland and El Salvador.

Iceland's Nesjavellir geothermal power station. Geothermal plants account for more than 25 percent of the electricity produced in Iceland.

With more than 3,300 megawatts in eight states, the United States is a global leader in installed geothermal capacity. Eighty percent of this capacity is located in California, where more than 40 geothermal plants provide nearly 7 percent of the state's electricity . In thousands of homes and buildings across the United States, geothermal heat pumps also use the steady temperatures just underground to heat and cool buildings, cleanly and inexpensively.

The Geothermal Resource

Below Earth's crust, there is a layer of hot and molten rock, called magma. Heat is continually produced in this layer, mostly from the decay of naturally radioactive materials such as uranium and potassium. The amount of heat within 10,000 meters (about 33,000 feet) of Earth's surface contains 50,000 times more energy than all the oil and natural gas resources in the world.

The areas with the highest underground temperatures are in regions with active or geologically young volcanoes. These "hot spots" occur at tectonic plate boundaries or at places where the crust is thin enough to let the heat through. The Pacific Rim, often called the Ring of Fire for its many volcanoes, has many hot spots, including some in Alaska, California, and Oregon. Nevada has hundreds of hot spots, covering much of the northern part of the state.

These regions are also seismically active. Earthquakes and magma movement break up the rock covering, allowing water to circulate. As the water rises to the surface, natural hot springs and geysers occur, such as Old Faithful at Yellowstone National Park. The water in these systems can be more than 200 °C (430 °F).

Seismically active hotspots are not the only places where geothermal energy can be found. There is a steady supply of milder heat—useful for direct heating purposes—at depths of anywhere from 10 to a few hundred feet below the surface virtually in any location on Earth. Even the ground below your own backyard or local school has enough heat to control the climate in your home or other buildings in the community. In addition, there is a vast amount of heat energy available from dry rock formations very deep below the surface (4–10 km). Using the emerging technology known as Enhanced Geothermal Systems (EGS), we may be able to capture this heat for electricity production on a much larger scale than conventional technologies currently allow. While still primarily in the development phase, the first demonstration EGS projects provided electricity to grids in the United States and Australia in 2013.

If the full economic potential of geothermal resources can be realized, they would represent an enormous source of electricity production capacity. In 2012, the U.S. National Renewable Energy Laboratory (NREL) found that conventional geothermal sources (hydrothermal) in 13 states have a potential capacity of 38,000 MW, which could produce 308 million MWh of electricity annually.

State and federal policies are likely to spur developers to tap some of this potential in the next few years. The Geothermal Energy Association estimates that 125 projects now under development around the country could provide up to 2,500 megawatts of new capacity.

As EGS technologies improve and become competitive, even more of the largely untapped geothermal resource could be developed. The NREL study found that hot dry rock resources could provide another 4 million MW of capacity, which is equivalent to more than all of today's U.S. electricity needs.

Not only do geothermal resources in the United States offer great potential, they can also provide continuous baseload electricity. According to NREL, the capacity factors of geothermal plants—a measure of the ratio of the actual electricity generated over time compared to what would be produced if the plant was running nonstop for that period—are comparable with those of coal and nuclear power. With the combination of both the size of the resource base and its consistency, geothermal can play an indispensable role in a cleaner, more sustainable power system.

Salt Wells geothermal plant in Nevada.

How Geothermal Energy is Captured

Geothermal springs for power plants. Currently, the most common way of capturing the energy from geothermal sources is to tap into naturally occurring "hydrothermal convection" systems, where cooler water seeps into Earth's crust, is heated up, and then rises to the surface. Once this heated water is forced to the surface, it is a relatively simple matter to capture that steam and use it to drive electric generators. Geothermal power plants drill their own holes into the rock to more effectively capture the steam.

There are three basic designs for geothermal power plants, all of which pull hot water and steam from the ground, use it, and then return it as warm water to prolong the life of the heat source. In the simplest design, known as dry steam, the steam goes directly through the turbine, then into a condenser where the steam is condensed into water. In a second approach, very hot water is depressurized or "flashed" into steam which can then be used to drive the turbine.

In the third approach, called a binary cycle system, the hot water is passed through a heat exchanger, where it heats a second liquid—such as isobutane—in a closed loop. Isobutane boils at a lower temperature than water, so it is more easily converted into steam to run the turbine. These three systems are shown in the diagrams below.

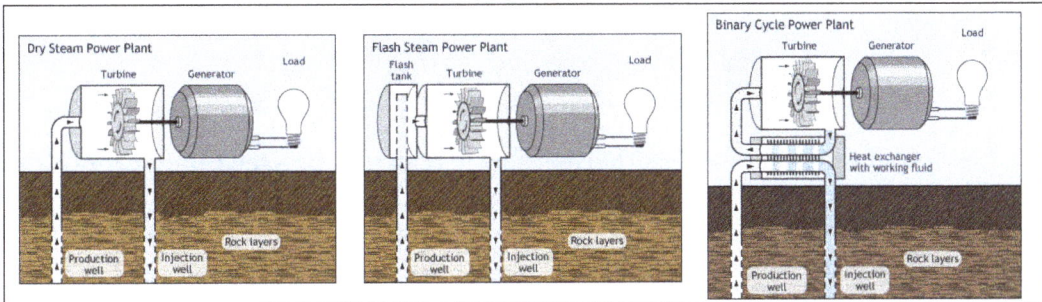

The three basic designs for geothermal power plants: dry steam, flash steam, and binary cycle.

The choice of which design to use is determined by the resource. If the water comes out of the well as steam, it can be used directly, as in the first design. If it is hot water of a high enough temperature, a flash system can be used; otherwise it must go through a heat exchanger. Since there are more hot water resources than pure steam or high-temperature water sources, there is more growth potential in the binary cycle, heat exchanger design.

The largest geothermal system now in operation is a steam-driven plant in an area called the Geysers, north of San Francisco, California. Despite the name, there are actually no geysers there, and the heat that is used for energy is all steam, not hot water. The area was known for its hot springs as far back as the mid-1800s, the first well for power production was not drilled until 1924. Deeper wells were drilled in the 1950s, but real development didn't occur until the 1970s and 1980s. By 1990, 26 power plants had been built, for a capacity of more than 2,000 MW.

Because of the rapid development of the area in the 1980s, and the technology used, the steam resource has been declining since 1988. Today, owned primarily by the California utility Calpine and with a net operating capacity of 725 MW, the Geysers facilities still meets nearly 60 percent of the average electrical demand for California's North Coast region (from the Golden Gate Bridge north to the Oregon border). The plants at the Geysers use an evaporative water-cooling process to create a vacuum that pulls the steam through the turbine, producing power more efficiently. But this process loses 60 to 80 percent of the steam to the air, without re-injecting it underground. While the steam pressure may be declining, the rocks underground are still hot. To remedy the situation, various stakeholders partnered to create the Santa Rosa Geysers Recharge Project, which involves transporting 11 million gallons per day of treated wastewater from neighboring communities through a 40-mile pipeline and injecting it into the ground to provide more steam. The project came online in 2003, and in 2008 provided enough additional electricity for approximately 100,000 homes.

One concern with open systems like the Geysers is that they emit some air pollutants. Hydrogen sulfide—a toxic gas with a highly recognizable "rotten egg" odor—along with trace amounts of arsenic and minerals, is released in the steam. Salt can also pose an environmental problem. At a power plant located at the Salton Sea reservoir in Southern California, a significant amount of salt builds up in the pipes and must be removed. While the plant initially put the salts into a landfill, they now re-inject the salt back into a different well. With closed-loop systems, such as the binary cycle system, there are no emissions and everything brought to the surface is returned underground.

Direct use of geothermal heat. Geothermal springs can also be used directly for heating purposes. Geothermal hot water is used to heat buildings, raise plants in greenhouses, dry out fish and crops, de-ice roads, improve oil recovery, aid in industrial processes like pasteurizing milk, and heat spas and water at fish farms. In Klamath Falls, Oregon, and Boise, Idaho, geothermal water has been used to heat homes and buildings for more than a century. On the east coast, the town of Warm Springs, Virginia obtains heat directly from spring water as well, using springs to heat one of the local resorts.

In Iceland, virtually every building in the country is heated with hot spring water. In fact, Iceland gets more than 50 percent of its primary energy from geothermal sources. In Reykjavik, for example (population 118,000), hot water is piped in from 25 kilometers away, and residents use it for heating and for hot tap water.

New Zealand's Wairakei geothermal power station.

Ground-source heat pumps. A much more conventional way to tap geothermal energy is by using geothermal heat pumps to provide heat and cooling to buildings. Also called ground-source heat pumps, they take advantage of the constant year-round temperature of about 50 °F that is just a few feet below the ground's surface. Either air or antifreeze liquid is pumped through pipes that are buried underground, and re-circulated into the building. In the summer, the liquid moves heat from the building into the ground. In the winter, it does the opposite, providing pre-warmed air and water to the heating system of the building.

In the simplest use of ground-source heating and cooling, a tube runs from the outside air, under the ground, and into a building's ventilation system. More complicated, but more effective, systems use compressors and pumps—as in electric air conditioning systems—to maximize the heat transfer.

In regions with temperature extremes, such as the northern United States in the winter and the southern United States in the summer, ground-source heat pumps are the most energy-efficient and environmentally clean heating and cooling systems available. Far more efficient than electric heating and cooling, these systems can circulate as much as 3 to 5 times the energy they use in the process. The U.S. Department of Energy found that heat pumps can save a typical home hundreds of dollars in energy costs each year, with the system typically paying for itself in 8 to 12 years. Tax credits and other incentives can reduce the payback period to 5 years or less.

More than 600,000 ground-source heat pumps supply climate control in U.S. homes and other buildings, with new installations occurring at a rate of about 60,000 per year. While this is significant, it is still only a small fraction of the U.S. heating and cooling market, and several barriers to greater penetration into the market remain. For example, despite their long-term savings, geothermal heat pumps have higher up-front costs. In addition, installing them in existing homes and businesses can be difficult, since it involves digging up areas around a building's structure. Finally, many heating and cooling installers are simply not familiar with the technology.

However, ground-source heat pumps are catching on in some areas. In rural areas without access to natural gas pipelines, homes must use propane or electricity for heating and cooling. Heat pumps are much less expensive to operate than these conventional systems, and since buildings are generally widely spread out, installing underground loops is often not an issue. Underground loops can be easily installed during construction of new buildings as well, resulting in savings for the life of the building. Furthermore, recent policy developments are offering strong incentives for homeowners to install these systems. The 2008 economic stimulus bill, Emergency Economic Stabilization Act of 2008, included an eight-year extension (through 2016) of the 30 percent investment tax credit, with no upper limit, to all home installations of EnergyStar certified geothermal heat pumps.

Geothermal energy has the potential to play a significant role in moving the United States (and other regions of the world) toward a cleaner, more sustainable energy system. It is one of the few renewable energy technologies that can supply continuous, baseload power. Additionally, unlike coal and nuclear plants, binary geothermal plants can be used a flexible source of energy to balance the variable supply of renewable resources such as wind and solar. Binary plants have the capability to ramp production up and down multiple times each day, from 100 percent of nominal power down to a minimum of 10 percent.

The costs for electricity from geothermal facilities are also becoming increasingly competitive. The U.S. Energy Information Administration (EIA) projected that the levelized cost of energy (LCOE) for new geothermal plants will be less than 5 cents per kilowatt hour (kWh), as opposed to more than 6 cents for new natural gas plants and more than 9 cents for new conventional coal. There is also a bright future for the direct use of geothermal resources as a heating source for homes and businesses in any location.

However, in order to tap into the full potential of geothermal energy, two emerging technologies require further development: Enhanced Geothermal Systems (EGS) and co-production of geothermal electricity in oil and gas wells.

Enhanced geothermal systems. Geothermal heat occurs everywhere under the surface of the earth, but the conditions that make water circulate to the surface are found in less than 10 percent of Earth's land area. An approach to capturing the heat in dry areas is known as enhanced geothermal systems (EGS) or "hot dry rock". The hot rock reservoirs, typically at greater depths below the surface than conventional sources, are first broken up by pumping high-pressure water through them. The plants then pump more water through the broken hot rocks, where it heats up, returns to the surface as steam, and powers turbines to generate electricity. The water is then returned to the reservoir through injection wells to complete the circulation loop. Plants that use a closed-loop binary cycle release no fluids or heat-trapping emissions other than water vapor, which may be used for cooling.

EGS technology could provide 100 gigawatts of electricity by 2050. The Department of Energy, several universities, the geothermal industry, and venture capital firms are collaborating on research and demonstration projects to harness the potential of EGS. The Newberry Geothermal Project in Bend, Oregon has recently made significant progress in reducing EGS project costs and eliminating risks to future development. The DOE hopes to have EGS ready for commercial development by 2015. Australia, France, Germany, and Japan also have R&D programs to make EGS commercially viable.

One cause for careful consideration with EGS is the possibility of induced seismic activity that might occur from hot dry rock drilling and development. This risk is similar to that associated with hydraulic fracturing, an increasingly used method of oil and gas drilling, and with carbon dioxide capture and storage in deep saline aquifers. Though a potentially serious concern, the risk of an induced EGS-related seismic event that can be felt by the surrounding population or that might cause significant damage currently appears very low when projects are located an appropriate distance away from major fault lines and properly monitored. Appropriate site selection, assessment and monitoring of rock fracturing and seismic activity during and after construction, and open, transparent communication with local communities are also critical.

Low-temperature and co-production of geothermal electricity in oil and gas wells. Low-temperature geothermal energy is derived from geothermal fluid found in the ground at temperatures of 150 °C (300 °F) or less. These resources are typically utilized in direct-use applications, such as heating buildings, but can also be used to produce electricity through binary cycle geothermal processes. Oil and gas fields already under production represent a large potential source of this type of geothermal energy. In many existing oil and gas reservoirs, a significant amount of high-temperature water or suitable high-pressure conditions are present, which could allow for the co-production of geothermal electricity along with the extraction of oil and gas resources. In some cases, exploiting these geothermal resources could even enhance the extraction of the oil and gas.

An MIT study estimated that the United States has the potential to develop 44,000 MWs of geothermal capacity by 2050 by coproducing geothermal electricity at oil and gas fields—primarily in the Southeast and southern Plains states. The study projected that such advanced geothermal systems could supply 10 percent of U.S. baseload electricity by 2050, given R&D and deployment over the next 10 years.

An average of 25 billion barrels of hot water is produced in United States oil and gas wells each year. This water, which has historically been viewed as an inconvenience to well operators, could be harnessed to produce up to 3 gigawatts of clean, reliable baseload energy. This energy could not only reduce greenhouse gas emissions, it could also increase profitability and extend the economic life of existing oil and gas field infrastructure.

GEOTHERMAL RESOURCES

Geothermal energy is present even below the coldest parts of Earth's surface — if you dig deep enough. In most places, Earth's heat is trapped at depths of more than 20 miles. But in so-called geological hotspots, geothermal energy is near the surface — and is apparent in geysers, hot springs and volcanic eruptions.

People have exploited geothermal energy for tens of thousands of years, using it to cook and bathe, for example. During the Roman Empire, about 2,000 years ago, hot springs were used for public baths and underfloor heating — as in the famous spa town of Bath, England.

The Blue Lagoon, a geothermal spa, is one of the most popular attractions in Iceland.

Today, geothermal energy tapped via holes drilled in the ground is used to heat and cool houses and other buildings. The underground environment functions as a sort of heat reservoir, with heat being drawn up into a building during cold weather and excess heat being dumped underground to lower indoor temperatures during hot weather.

The cost of a geothermal heat pump system varies according to the climate and other factors. A typical residential system might be twice that of a conventional heating and cooling system, but geothermal heating and cooling systems can cut utility costs by up to 60 percent.

Geothermal heat is also used in certain industrial and agricultural processes — for example, to dry lumber and crops.

Electricity from Underground Heat

Geothermal energy is also used to generate electricity. Geothermal power plants now generate about 0.5 percent of the electricity used in the U.S. which is the world's leading producer of geothermal electricity. In California, along the geothermally rich region known as the Pacific Ring of Fire, geothermal plants now provide more than 5 percent of the state's electricity.

Geothermal power plants typically draw energy from so-called production wells drilled to depths ranging from 500 feet to two miles. Steam and superheated water at these depths rise under their own pressure to turn electricity-generating turbines at the surface; waste liquid from the process is captured and returned underground through what are called injection wells.

Nonstop Energy

Geothermal power plants cost more to build than typical natural gas power plants, but the cost of operating a geothermal plant is usually far lower. That's mainly because geothermal power plants don't require fuel. Fuel costs for a power plant that uses natural gas, oil or coal can be double the cost of building the station itself.

Electricity generated by geothermal plants is often less expensive than electricity generated by wind, hydro power and solar.

GEOTHERMAL EXPLORATION

Geothermal exploration is the exploration of the subsurface in search of viable active geothermal regions with the goal of building a geothermal power plant, where hot fluids drive turbines to create electricity. Exploration methods include a broad range of disciplines including geology, geophysics, geochemistry and engineering.

Geothermal regions with adequate heat flow to fuel power plants are found in rift zones, subduction zones and mantle plumes. Hot spots are characterized by four geothermal elements. An active region will have:

1. Heat Source: Shallow magmatic body, decaying radioactive elements or ambient heat from high pressures.

2. Reservoir: Collection of hot rocks from which heat can be drawn.

3. Geothermal Fluid: Gas, vapor and water found within the reservoir.

4. Recharge Area: Area surrounding the reservoir that rehydrates the geothermal system.

Exploration involves not only identifying hot geothermal bodies, but also low-density, cost effective regions to drill and already constituted plumbing systems inherent within the subsurface. This information allows for higher success rates in geothermal plant production as well as lower drilling costs.

As much as 42% of all expenses associated with geothermal energy production can be attributed to exploration. These costs are mostly from drilling operations necessary to confirm or deny viable geothermal regions. Some geothermal experts have gone to say that developments in exploration techniques and technologies have the potential to bring the greatest advancements within the industry.

Methods of Exploration

Drilling

Drilling provides the most accurate information in the exploration process, but is also the most costly exploration method.

Thermal gradient holes (TGH), exploration wells (slim holes), and full-scale production wells (wildcats) provide the most reliable information on the subsurface. Temperature gradients, thermal pockets and other geothermal characteristics can be measured directly after drilling, providing valuable information.

Geothermal exploration wells rarely exceed 4 km in depth. Subsurface materials associated with geothermal fields range from limestone to shale, volcanic rocks and granite. Most drilled geothermal exploration wells, up to the production well, are still considered to be within the exploration phase. Most consultants and engineers consider exploration to continue until one production well is completed successfully.

Generally, the first wildcat well has a success rate of 25%. Following more analysis and investigation, success rates then increase to a range from 60% to 80%. Although expenses vary significantly, drilling costs are estimated at $400/ft. Therefore, it is becoming paramount to investigate other means of exploration before drilling operations commence. To increase the chances of successfully drilling, innovations in remote sensing technologies have developed over the last 2 decades. These less costly means of exploration are categorized into multiple fields including geology, geochemistry and geophysics.

Geophysics

Seismology

Seismology has played a significant role in the oil and gas industry and is now being adapted to geothermal exploration. Seismic waves propagate and interact with subterranean components and respond accordingly. Two sub categories exist that are relevant to the source of the seismic signal. Active seismology relies on using induced/man-made vibrations at or near the surface. Passive seismology uses earthquakes, volcanic eruptions or other tectonic activity as sources.

Passive seismic studies use natural wave propagation through the earth. Geothermal fields are often characterized by increased levels of seismicity. Earthquakes of lesser magnitude are much more frequent than ones of larger magnitude. Therefore, these micro earthquakes (MEQ), registering below 2.0 magnitude on the Richter scale, are used to reveal subsurface qualities relating to geothermal exploration. The high rate of MEQ in geothermal regions produce large datasets that do not require long field deployments.

Active Seismology, which has history in the oil and gas industry, involves studying man made vibrational wave propagation. In these studies geophones (or other seismic sensors) are spread across the study site. The most common geophone spreads are in line, offset, in-line with center shot and Fan shooting.

Many analytical techniques can be applied to active seismology studies but generally all include Huygens Principle, Fermat's Principle and Snell's law. These basic principles can be used to identify subsurface anomalies, reflective layers and other objects with high impedance contrasts.

Gravity

Gravimetry studies use changes in densities to characterize subsurface properties. This method is well applied when identifying dense subsurface anomalies including granite bodies, which are vital to locate in the geothermal exploration projects. Subsurface fault lines are also identifiable with gravitational methods. These faults are often identified as prime drilling locations as their densities are much less than surrounding material. Developments in airborne gravitational studies yield large amounts of data, which can be used to model the subsurface 3 dimensionally with relatively high levels of accuracy.

Changes in groundwater levels may also be measured and identified with gravitational methods. This recharge element is imperative in creating productive geothermal systems. Pore density and subsequent overall density are affected by fluid flow and therefore change the gravitational field. When correlated with current weather conditions, this can be measured and modeled to estimate the rate of recharge in geothermal reservoirs.

Unfortunately, there are many other factors that must be realized before data from a gravity study can be interpreted. The average gravitational field the earth produces is 920 cm². Objects of concern produce a significantly smaller gravitational field. Therefore, instrumentation must detect variations as small as 0.00001%. Other considerations including elevation, latitude and weather conditions must be carefully observed and taken into account.

Resistivity and Magnetotellurics

Magnetotellurics (MT) measurements allow detection of resistivity anomalies associated with productive geothermal structures, including faults and the presence of a cap rock, and allow for estimation of geothermal reservoir temperatures at various depths. MT has successfully contributed to the successful mapping and development of geothermal resources around the world since the early 1980s, including in the U.S. and countries located on the Pacific Ring of Fire such as Japan, New Zealand, the Philippines, Ecuador, and Peru.

Geological materials are generally poor electrical conductors and have a high resistivity. Hydrothermal fluids in the pores and fractures of the earth, however, increase the conductivity of the subsurface material. This change in conductivity is used to map the subsurface geology and estimate the subsurface material composition. Resistivity measurements are made using a series of probes distributed tens to hundreds of meters apart, to detect the electrical response of the Earth to injection of electrical impulses in order to reconstruct the distribution of electrical resistance in the rocks. Since flowing geothermal waters can be detected as zones of low resistance, it is possible to map geothermal resources using such a technique. However, care must be exercised when interpreting low resistivity zones since they may also be caused by changes in rock type and temperature.

The Earth's magnetic field varies in intensity and orientation during the day inducing detectable electrical currents in the Earth's crust. The range of the frequency of those currents allows a multispectral analysis of the variation in the electromagnetic local field. As a result, it is possible a tomographic reconstruction of geology, since the currents are determined by the underlying response of the different rocks to the changing magnetic field.

Magnetics

The most common application magnetism has in geothermal exploration involves identifying the depth of the curie point or curie temperature. At the curie point, materials will change from ferromagnetic to paramagnetic. Locating curie temperatures for known subsurface materials provides estimates on future plant productivity. For example, titanomagnetitite, a common material in geothermal fields, has a curie temperature between 200-570 degrees Celsius. Simple geometric anomalies modeled at different depths are used to best estimate the curie depth.

Geochemistry

This science is readily used in geothermal exploration. Scientists within this field relate surface fluid properties and geologic data to geothermal bodies. Temperature, isotopic ratios, elemental ratios, mercury & CO_2 concentrations are all data points under close examination. Geothermometers and other instrumentation are placed around field sites to increase the fidelity of subsurface temperature estimates.

HYDROTHERMAL EXPANSION

Hydrothermal explosions occur when superheated water trapped below the surface of the earth rapidly converts from liquid to steam, violently disrupting the confining rock. Boiling water, steam, mud, and rock fragments called breccia are ejected over an area of a few meters up to several kilometers in diameter. Although the energy inherently comes from a deep igneous source, this energy is transferred to the surface by circulating meteoric water rather than by magma, as occurs in volcanic eruptions. The energy is stored as heat in hot water and rock within a few hundred feet of the surface.

Hydrothermal explosions are caused by the same instability and chain reaction mechanism as geysers but are so violent that rocks and mud are expelled along with water and steam.

Cause

Hydrothermal explosions occur where shallow interconnected reservoirs of water at temperatures as high as 250° Celsius underlie thermal fields. Water usually boils at 100 °C, but under pressure its boiling point increases, causing the water to become super-heated. A sudden reduction in pressure causes a rapid phase transition from liquid to steam, resulting in an explosion of water and rock debris. During the last Ice Age, many hydrothermal explosions were triggered by the release of pressure as glaciers receded. Other causes are seismic activity, erosion, or hydraulic fracturing.

Yellowstone

Yellowstone National Park is a thermally active area with an extensive system of hot springs, fumaroles, geysers, and mudpots. There are also several hydrothermal explosion craters, which are collapse features. Eight of these hydrothermal explosion craters are in hydrothermally cemented glacial deposits, and two are in Pleistocene ash-flow tuff. Each is surrounded by a rim composed of debris derived from the crater, 30 to 100 feet high.

More than 20 large hydrothermal explosions have occurred at Yellowstone, approximately one every 700 years. The temperature of the magma reservoir below Yellowstone is believed to exceed 800° Celsius causing the heating of rocks in the region. If so, the average heat flow supplied by convection currents is 30 times greater than anywhere in the Rocky Mountains. Snowmelt and rainfall seep into the ground at a rapid rate and can conduct enough heat to raise the temperature of ground water to almost boiling.

The phenomena of geyser basins are the product of hot ground water rising close to the surface and occasionally bubbling through. Water temperatures of 238° Celsius at 332 meters have been recorded at Norris Geyser Basin. Pocket Basin was originally an ice-dammed lake over a hydrothermal system. Melting ice during the last glacial period caused the lake to rapidly drain, causing a sudden change in pressure triggering a massive hydrothermal explosion.

Geysers

A hydrothermal explosion is similar to a geyser's eruption except that includes surrounding rock and mud.

One well-known hydrothermal geyser is Old Faithful which throws up plumes of steam and water approximately every hour and a half on average. Rarely has any steam explosion violently hurled water and rock thousands of feet above the ground; however in Yellowstone's geological history these colossal events have been recorded numerous times and have been found to have created new hills and shaped parts of the landscape.

The largest hydrothermal explosion ever documented was located near the northern edge of Yellowstone Lake, on an embankment commonly known as "Mary Bay". Now consisting of a 1.5 mile crater, it was formed relatively recently, approximately 13,800 years ago. It is believed this crater was formed by a sequence of several hydrothermal explosions in a short time. What triggered this series of events has not yet been clearly established, but volcanologists believe a large earthquake could have played a role by accelerating the melting of nearby glaciers and thus depressurizing the hydrothermal system. Alternatively, rapid changes in the level of Yellowstone Lake may have been responsible.

Recent Explosions

Most of Yellowstone's recent large hydrothermal explosions have been the consequence of sudden changes of pressure deep within the hydrothermal system. Generally, these larger explosions have created craters in a north-south pattern (between Norris and Mammoth Hot Springs). It is estimated that all of the known hydrothermal craters were created between 14,000 and 3,000 years ago. Volcanologists believe no magma has ever broken through the fragile crust of Yellowstone Park or stirred the movement of magma in the reservoir beneath Yellowstone. These phenomena are now considered to be mutually exclusive events; hydrothermal explosions are not correlated with volcanism, although throughout the world all hydrothermal systems are heated and caused by magma.

GEOTHERMAL HEATING AND COOLING TECHNOLOGIES

Geothermal technology harnesses the Earth's heat. Just a few feet below the surface, the Earth maintains a near-constant temperature, in contrast to the summer and winter extremes of the ambient air above ground. Farther below the surface, the temperature increases at an average rate of approximately 1°F for every 70 feet in depth. In some regions, tectonic and volcanic activity can bring higher temperatures and pockets of superheated water and steam much closer to the surface.

Three main types of technologies take advantage of Earth as a heat source:

- Ground source heat pumps.

- Direct use geothermal.

- Deep and enhanced geothermal systems.

Geothermal energy is considered a renewable resource. Ground source heat pumps and direct use geothermal technologies serve heating and cooling applications, while deep and enhanced geothermal technologies generally take advantage of a much deeper, higher temperature geothermal resource to generate electricity.

Ground Source Heat Pumps

A ground source heat pump takes advantage of the naturally occurring difference between the above-ground air temperature and the subsurface soil temperature to move heat in support of end uses such as space heating, space cooling (air conditioning), and even water heating. A ground source or geoexchange system consists of a heat pump connected to a series of buried pipes. One can install the pipes either in horizontal trenches just below the ground surface or in vertical boreholes that go several hundred feet below ground. The heat pump circulates a heat-conveying fluid, sometimes water, through the pipes to move heat from point to point.

A commercial-scale ground source heat pump system. This example is a demonstration project at a university.

If the ground temperature is warmer than the ambient air temperature, the heat pump can move heat from the ground to the building. The heat pump can also operate in reverse, moving heat from the ambient air in a building into the ground, in effect cooling the building. Ground source heat pumps require a small amount of electricity to drive the heating/cooling process. For every unit of electricity used in operating the system, the heat pump can deliver as much as five times the energy from the ground, resulting in a net energy benefit. Geothermal heat pump users should be aware that in the absence of using renewable generated electricity to drive the heating/cooling process (e.g. modes) that geothermal heat pump systems may not be fully fossil-fuel free (e.g. renewable-based).

How it Works

The steps below describe how a heat pump works in "heating mode"—taking heat from the ground and delivering it to a building—and "cooling mode," which removes heat from the building and transfers it to the ground.

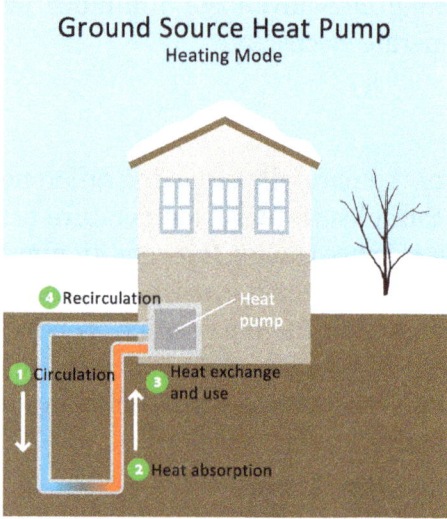

Heating Mode.

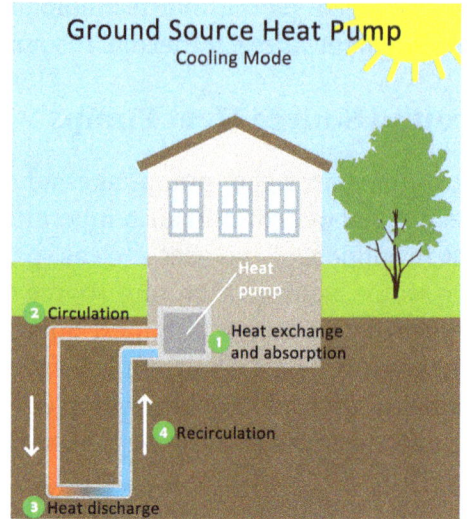

Cooling Mode.

1. Circulation: The above-ground heat pump moves water or another fluid through a series of buried pipes or ground loops.

2. Heat absorption: As the fluid passes through the ground loop, it absorbs heat from the warmer soil, rock, or ground water around it.

3. Heat exchange and use: The heated fluid returns to the building where it used for useful purposes, such as space or water heating. The system uses a heat exchanger to transfer heat into the building's existing air handling, distribution, and ventilation system, or with the addition of a desuperheater it can also heat domestic water.

4. Recirculation: Once the fluid transfers its heat to the building, it returns at a lower temperature to the ground loop to be heated again. This process is repeated, moving heat from one point to another for the user's benefit and comfort.

1. Heat exchange and absorption: Water or another fluid absorbs heat from the air inside the building through a heat exchanger, which is the way a typical air conditioner works.

2. Circulation: The above-ground heat pump moves the heated fluid through a series of buried pipes or ground loops.

3. Heat discharge: As the heated fluid passes through the ground loop, it gives off heat to the relatively colder soil, rock, or ground water around it.

4. Recirculation: Once the fluid transfers its heat to the ground, the fluid returns at a lower temperature to the building, where it absorbs heat again. This process is repeated, moving heat from one point to another for the user's benefit and comfort.

The above-ground heat pump is relatively inexpensive, with underground installation of ground loops (piping) accounting for most of the system's cost. Heat pumps can support space heating and cooling needs in almost any part of the country, and they can also be used for domestic hot water applications. Increasing the capacity of the piping loops can scale this technology for larger buildings or locations where space heating and cooling, as well as water heating, may be needed for most of the year.

Direct use Geothermal

Direct use geothermal systems use groundwater that is heated by natural geological processes below the Earth's surface. This water can be as hot as 200°F or more. Bodies of hot groundwater can be found in many areas with volcanic or tectonic activity. In locations such as Yellowstone National Park and Iceland, these groundwater reservoirs can reach the surface, creating geysers and hot springs. One can pump hot water from the surface or from underground for a wide range of useful applications.

Geothermally heated water reaches the surface at hot springs like this one in Yellowstone National Park.

1. Pumping: To tap into hot ground water, a well is drilled. A pumping system may be installed, although in some cases, hot water or steam may rise up through the well without active pumping.

2. Delivery: Hot water or steam can be used directly in a variety of applications, or it can be cycled through a heat exchanger.

3. Refilling: Depending on the use requirements of the system and the conditions of the site, the ground water aquifer may need to be replenished with water from the surface. In some cases, the movement of ground water might refill the aquifer naturally.

The water from direct geothermal systems is hot enough for many applications, including large-scale pool heating; space heating, cooling, and on-demand hot water for buildings of most sizes; district heating (i.e. heat for multiple buildings in a city); heating roads and sidewalks to melt snow; and some industrial and agricultural processes.

Direct use takes advantage of hot water that may be just a few feet below the surface, and usually less than a mile deep. The shallow depth means that capital costs are relatively small compared with deeper geothermal systems, but this technology is limited to regions with natural sources of hot groundwater at or near the surface.

Deep and Enhanced Geothermal Systems

Deep geothermal systems use steam from far below the Earth's surface for applications that require temperatures of several hundred degrees Fahrenheit. These systems typically inject water into the ground through one well and bring water or steam to the surface through another. Other variations can capture steam directly from underground ("dry steam"). Unlike ground source heat pumps or direct use geothermal systems, deep geothermal projects can involve drilling a mile or more below the Earth's surface. At these depths, high pressure keeps the water in a liquid state even at temperatures of several hundred degrees Fahrenheit.

How it Works

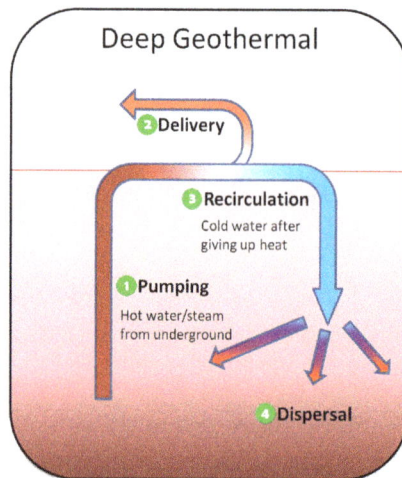

1. Pumping: Hot water or steam is pumped up through a deep well. As the water rises to the surface, the pressure drops and the water vaporizes into superheated steam that can be used for high-temperature processes.

2. Delivery: The heat from the hot water or steam can be used to heat a secondary fluid (a "binary" process), or the hot water or steam can be used directly.

3. Recirculation: Once the heat is transferred to the delivery system, the now-cooler water is pumped back underground.

4. Dispersal: Unlike ground source heat pumps, used ground water in this case is simply injected and allowed to disperse back into the ground, rather than being pumped through a closed loop of pipes.

Deep geothermal technologies harness the same kind of energy that produces geysers.

Deep geothermal sources provide efficient, clean heat for industrial processes and some large-scale commercial and agricultural uses. In addition, steam can be used to spin a turbine and generate electricity. Although geothermal steam requires no fuel and low operational costs, the initial capital costs—especially drilling test wells and production wells—can be financially challenging. Steam resources that are economical to tap into are currently limited to regions with high geothermal activity, but research is underway to develop enhanced geothermal systems with much deeper wells that take advantage of the Earth's natural temperature gradient and can potentially be constructed any-where. Enhanced systems can use hydraulic fracturing techniques to engineer subsur-face reservoirs that allow water to be pumped into and through otherwise dry or im-permeable rock.

References

- Geothermal-energy, science: britannica.com, Retrieved 30 April, 2019

- Rose, william ingersoll (2004), natural hazards in el salvador, geological society of america, pp. 246–247, isbn 0-8137-2375-2

- How-geothermal-energy-works, renewable-energy, our-energy-choices, clean-energy: ucsusa.org, Retrieved 29 June, 2019

- What-geothermal-energy-ncna, science, mach: nbcnews.com, Retrieved 14 July, 2019

- Robert baer smith, lee j. Siegel (2000), windows into the earth: the geologic story of yellowstone and grand teton national parks, oxford university press us, pp. 64–66, isbn 0-19-510597-4

- Geothermal-heating-and-cooling-technologies: epa.gov, Retrieved 11 January, 2019

5

Biomass Energy

The renewable energy which is derived from the organic material that is obtained from plants and animals is known as biomass energy. Biofuel is the most common fuel that is derived from biomass. This chapter closely examines the varied uses of biomass energy to provide an extensive understanding of the subject.

Biomass is organic material that comes from plants and animals, and it is a renewable source of energy.

Biomass contains stored energy from the sun. Plants absorb the sun's energy in a process called photosynthesis. When biomass is burned, the chemical energy in biomass is released as heat. Biomass can be burned directly or converted to liquid biofuels or biogas that can be burned as fuels.

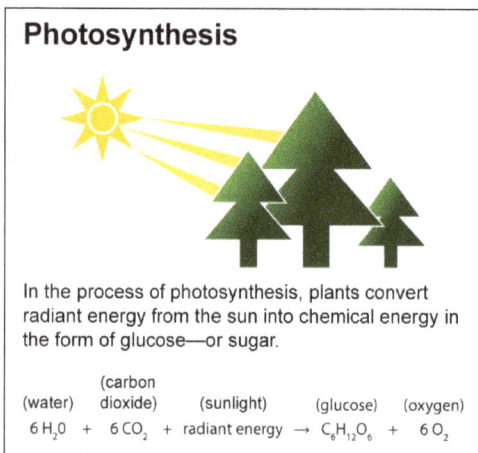

Photosynthesis

In the process of photosynthesis, plants convert radiant energy from the sun into chemical energy in the form of glucose—or sugar.

$$\underset{\text{(water)}}{6\,H_2O} \;+\; \underset{\substack{\text{(carbon}\\ \text{dioxide)}}}{6\,CO_2} \;+\; \text{radiant energy} \;\underset{\text{(sunlight)}}{} \;\rightarrow\; \underset{\text{(glucose)}}{C_6H_{12}O_6} \;+\; \underset{\text{(oxygen)}}{6\,O_2}$$

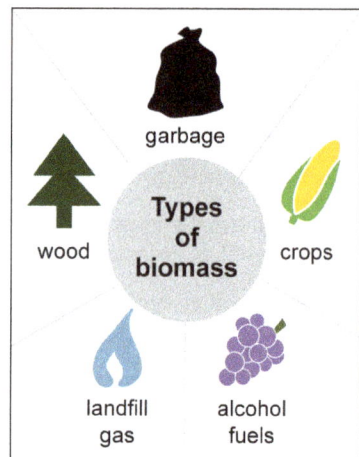

garbage

wood   **Types of biomass**   crops

landfill gas   alcohol fuels

Examples of biomass and their uses for energy:

- Wood and wood processing wastes: Burned to heat buildings, to produce process heat in industry, and to generate electricity.

- Agricultural crops and waste materials: Burned as a fuel or converted to liquid biofuels.

- Food, yard, and wood waste in garbage: Burned to generate electricity in power plants or converted to biogas in landfills.

- Animal manure and human sewage: Converted to biogas, which can be burned as a fuel.

Converting Biomass to Energy

Solid biomass, such as wood and garbage, can be burned directly to produce heat. Biomass can also be converted into a gas called biogas or into liquid biofuels such as ethanol and biodiesel. These fuels can then be burned for energy.

Biogas forms when paper, food scraps, and yard waste decompose in landfills, and it can be produced by processing sewage and animal manure in special vessels called digesters.

Ethanol is made from crops such as corn and sugar cane that are fermented to produce fuel ethanol for use in vehicles. Biodiesel is produced from vegetable oils and animal fats and can be used in vehicles and as heating oil.

How much Biomass is used for Fuel?

Biomass fuels provided about 5% of total primary energy use in the United States in 2017. Of that 5%, about 47% was from biofuels (mainly ethanol), 44% was from wood and wood-derived biomass, and 10% was from the biomass in municipal waste. (Sum of percentages is greater than 100% because of independent rounding) Researchers are trying to develop ways to use more biomass for fuel.

USES OF BIOMASS ENERGY

Biomass fuels come from things which once existed such as wood products, crop waste, garbage, dried vegetation and aquatic plants. Plants retrieved their energy from the sun during photosynthesis process.

This energy is stored in plants in form of chemical energy. Once the plant is dead, this energy is still trapped inside the plant. When biomass is burned, this energy releases as heat and produces energy.

Carbon dioxide is released while burning these plants, the same carbon dioxide which plants once used to grow their leaves and branches. Once the energy is released, the same carbon dioxide is returned to the air.

Biomass is a constant source of producing energy because:

- Plants will always exist if we keep on planting the new ones.
- Waste materials such as garbage leftover crops, scrap wood will always exist.
- Animal waste will also be available.

Biomass powerhouse is producing renewable energies by using organic matters which is being used for multiple purposes such as follows.

Production of Fuels

Biomass energy is produced through the process of fermentation. Yeast (an element of bacteria) is added to biomass waste such as wood and agricultural waste to produce ethanol. Ethanol is used in place gasoline to power cars. Biomass energy is available are three forms of fuel:

- Solid-compressed pieces of organic matter which release their energy through Ignition and burning.

- Liquid-fluid produced through organic matter which is used to fuel automobiles.

- Gas-a kind of natural extracted gas from decayed plants and dead animals, used in cars with the name.

Energy Generation

Biomass energy is also used to produce electricity. Powerhouses use heat and steam produced by burning organic matters to generate electricity.

However, most powerhouse uses fossil fuels (coal) to produce electricity. In powerhouses, they burn wood waste and other waste materials to produce steam that runs a turbine to produce electricity.2000 pounds of the garbage can produce as much energy as pounds of coal.

Currently, biomass is producing electricity which is being supplied to 1.3 million USA homes.

Thermal Burning

By burning solid biomass materials, we gain energy to fuel our homes and industries such as water heating, cooking, and washing. It is the most common and domesticated use of biomass energy in our lives.

High productivity home stoves and fireplace areas are also widely used. Large furnaces and boilers are used in industries by burning various types of waste materials.

Biomass is environmentally friendly and can be produced by utilizing waste materials which are of no use to anyone. It is better than burning fossil fuels which produce pollutants such as sulfur.

While growing plants, we can help save our environment as plants emit carbon dioxide to grow while releasing oxygen into the air.

BIOFUEL

A biofuel is a fuel that is produced through contemporary processes from biomass, rather than a fuel produced by the very slow geological processes involved in the formation of fossil fuels, such as oil. Since biomass technically can be used as a fuel directly (e.g. wood logs), some people use the terms biomass and biofuel interchangeably. More often than not however, the word biomass simply denotes the biological raw material the fuel is made of, or some form of thermally/chemically altered *solid* end product, like torrefied pellets or briquettes. The word biofuel is usually reserved for liquid or gaseous fuels, used for transportation. If the biomass used in the production of biofuel can regrow quickly, the fuel is generally considered to be a form of renewable energy.

Biofuel logo.

Biofuels can be produced from plants (i.e. energy crops), or from agricultural, commercial, domestic, and/or industrial wastes (if the waste has a biological origin). Renewable biofuels generally involve contemporary carbon fixation, such as those that occur in plants or microalgae through the process of photosynthesis.

Some argue that biofuel can be carbon-neutral because all biomass crops sequester carbon to a certain extent – basically all crops move CO_2 from above-ground circulation to below-ground storage in the roots and the surrounding soil. For instance, McCalmont et al. found below-ground carbon accumulation ranging from 0.42 to 3.8 tonnes per hectare per year for soils below Miscanthus x giganteus energy crops, with a mean accumulation rate of 1.84 tonne (0.74 tonnes per acre per year), or 20% of total harvested carbon per year.

However, the simple proposal that biofuel is carbon-neutral almost by definition has been superseded by the more nuanced proposal that for a particular biofuel project to be carbon neutral, the total carbon sequestered by the energy crop's root system must compensate for all the above-ground emissions (related to this particular biofuel project). This includes any emissions caused by direct or indirect land use change. Many first generation biofuel projects are not carbon neutral given these demands. Some have even higher total GHG emissions than some fossil based alternatives.

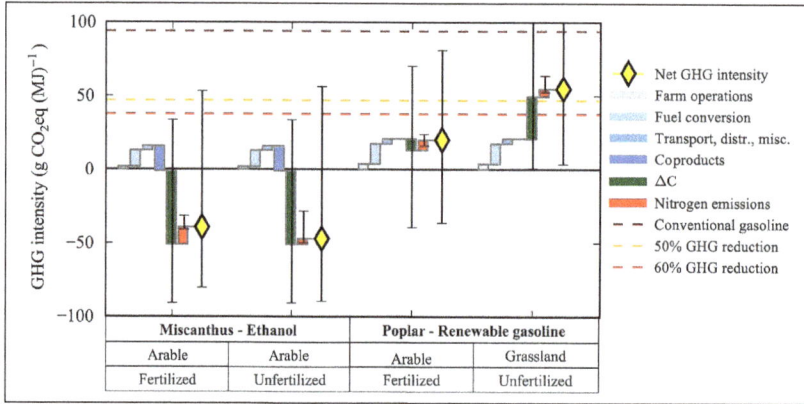

GHG / CO_2 / carbon negativity for Miscanthus x giganteus production pathways.

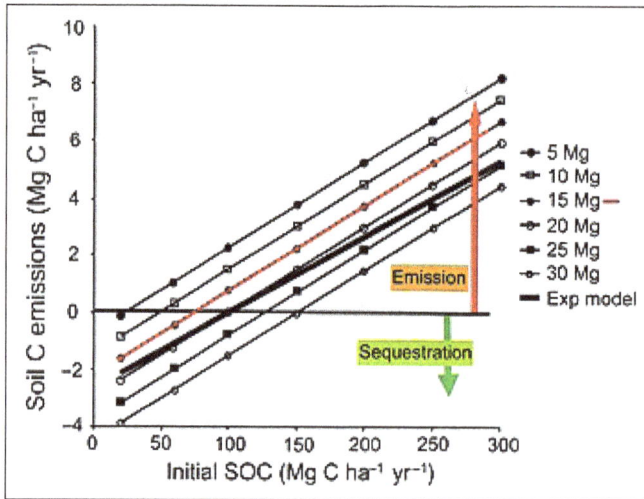

Relationship between above-ground yield (diagonal lines), soil organic carbon (X axis), and soil's potential for successful/unsuccessful carbon sequestration (Y axis). Basically, the higher the yield, the more land is usable as a GHG mitigation tool (including relatively carbon rich land.)

Some are carbon neutral or even negative, though, especially perennial crops. The amount of carbon sequestrated and the amount of GHG (greenhouse gases) emitted will determine if the total GHG life cycle cost of a biofuel project is positive, neutral or negative. A carbon negative life cycle is possible if the total below-ground carbon accumulation more than compensates for the total life-cycle GHG emissions above ground. In other words, to achieve carbon neutrality yields should be high and emissions should be low.

High-yielding energy crops are thus prime candidates for carbon neutrality. The graphic on the right displays two CO_2 negative Miscanthus x giganteus production pathways, represented in gram CO_2-equivalents per megajoule. The yellow diamonds represent mean values. Further, successful sequestration is dependent on planting sites, as the best soils for sequestration are those that are currently low in carbon. The varied results displayed in the graph highlights this fact. For the UK, successful sequestration is expected for arable land over most of England and Wales, with unsuccessful sequestration expected in parts of Scotland, due to already carbon rich soils (existing woodland) plus lower yields. Soils already rich in carbon includes peatland and mature forest. Grassland can also be carbon rich, however Milner et al. argues that the most successful carbon sequestration in the UK takes place below improved grasslands. The bottom graphic displays the estimated yield necessary to compensate for related lifecycle GHG-emissions. The higher the yield, the more likely CO_2 negativity becomes.

The two most common types of biofuel are bioethanol and biodiesel:

- Bioethanol is an alcohol made by fermentation, mostly from carbohydrates produced in sugar or starch crops such as corn, sugarcane, or sweet sorghum. Cellulosic biomass, derived from non-food sources, such as trees and grasses, is also being developed as a feedstock for ethanol production. Ethanol can be used as a fuel for vehicles in its pure form (E100), but it is usually used as a gasoline additive to increase octane and improve vehicle emissions. Bioethanol is widely used in the United States and in Brazil.

- Biodiesel is produced from oils or fats using transesterification and is the most common biofuel in Europe. It can be used as a fuel for vehicles in its pure form (B100), but it is usually used as a diesel additive to reduce levels of particulates, carbon monoxide, and hydrocarbons from diesel-powered vehicles.

In 2018, worldwide biofuel production reached 152 billion liters (40 billion gallons US), up 7% from 2017, and biofuels provided 3% of the world's fuels for road transport. The International Energy Agency want biofuels to meet more than a quarter of world demand for transportation fuels by 2050, in order to reduce dependency on petroleum. However, the production and consumption of biofuels are not on track to meet the IEA's sustainable development scenario. From 2020 to 2030 global biofuel output has to increase by 10% each year to reach IEA's goal. Only 3% growth annually is expected.

There are various social, economic, environmental and technical issues relating to biofuels production and use, which have been debated in the popular media and scientific journals.

Generations

First-Generation Biofuels

"First-generation" or conventional biofuels are biofuels made from food crops grown on arable land. With this biofuel production generation, food crops are thus explicitly grown for fuel production, and not anything else. The sugar, starch, or vegetable oil obtained from the crops is converted into biodiesel or ethanol, using transesterification, or yeast fermentation.

Second-generation Biofuels

Second generation biofuels are fuels manufactured from various types of biomass. Biomass is a wide-ranging term meaning any source of organic carbon that is renewed rapidly as part of the carbon cycle. Biomass is derived from plant materials, but can also include animal materials.

Whereas first generation biofuels are made from the sugars and vegetable oils found in arable crops, second generation biofuels are made from lignocellulosic biomass or woody crops, agricultural residues or waste plant material (from food crops that have already fulfilled their food purpose). The feedstock used to generate second-generation biofuels thus either grows on arable lands, but are just byproducts of the actual harvest (main crop) or they are grown on lands which cannot be used to effectively grow food crops and in some cases neither extra water or fertilizer is applied to them. Non-human food second generation feedstock sources include grasses, jatropha and other seed crops, waste vegetable oil, municipal solid waste and so forth.

This has both advantages and disadvantages. The advantage is that, unlike with regular food crops, no arable land is used solely for the production of fuel. The disadvantage is that unlike with regular food crops, it may be rather difficult to extract the fuel. For instance, a series of physical and chemical treatments might be required to convert lignocellulosic biomass to liquid fuels suitable for transportation.

Third-generation Biofuels

From 1978 to 1996, the US NREL experimented with using algae as a biofuels source in the Aquatic Species Program. The realistic replacement of all vehicular fuel with biofuels by using algae that have a natural oil content greater than 50%, which Briggs suggests can be grown on algae ponds at wastewater treatment plants. This oil-rich algae can then be extracted from the system and processed into biofuels, with the dried remainder further reprocessed to create ethanol. The production of algae to harvest oil for biofuels has not yet been undertaken on a commercial scale, but feasibility studies have been conducted to arrive at the above yield estimate. In addition to its projected high yield, algaculture – unlike crop-based biofuels – does not entail a decrease in food production, since it requires neither farmland nor fresh water. Many companies are pursuing algae bioreactors for various purposes, including scaling up biofuels production to commercial levels. Prof. Rodrigo E. Teixeira from the University of Alabama in Huntsville demonstrated the extraction of biofuels lipids from wet algae using a simple and economical reaction in ionic liquids.

Fourth-Generation Biofuels

Similarly to third-generation biofuels, fourth-generation biofuels are made using non-arable land. However, unlike third-generation biofuels, they do not require the destruction of biomass. This class of biofuels includes electrofuels and photobiological solar fuels. Some of these fuels are carbon-neutral.

Types

The following fuels can be produced using first, second, third or fourth-generation biofuel production procedures. Most of these can even be produced using two or three of the different biofuel generation procedures.

Biogas

Pipes carrying biogas.

Biogas is methane produced by the process of anaerobic digestion of organic material by anaerobes. It can be produced either from biodegradable waste materials or by the use of energy crops fed into anaerobic digesters to supplement gas yields. The solid byproduct, digestate, can be used as a biofuel or a fertilizer.

Biogas can be recovered from mechanical biological treatment waste processing systems. Landfill gas, a less clean form of biogas, is produced in landfills through naturally occurring anaerobic digestion. If it escapes into the atmosphere, it is a potential greenhouse gas. Farmers can produce biogas from manure from their cattle by using anaerobic digesters.

Syngas

Syngas, a mixture of carbon monoxide, hydrogen and other hydrocarbons, is produced by partial combustion of biomass, that is, combustion with an amount of oxygen that is not sufficient to convert the biomass completely to carbon dioxide and water. Before partial combustion, the biomass is dried, and sometimes pyrolysed. The resulting gas mixture, syngas, is more efficient than direct combustion of the original biofuel; more of the energy contained in the fuel is extracted.

Syngas may be burned directly in internal combustion engines, turbines or high-temperature fuel cells. The wood gas generator, a wood-fueled gasification reactor, can be connected to an internal combustion engine.

Syngas can be used to produce methanol, DME and hydrogen, or converted via the Fischer-Tropsch process to produce a diesel substitute, or a mixture of alcohols that can be blended into gasoline. Gasification normally relies on temperatures greater than 700 °C.

Lower-temperature gasification is desirable when co-producing biochar, but results in syngas polluted with tar.

Ethanol

Neat ethanol on the left (A), gasoline on the right (G) at a filling station in Brazil.

Biologically produced alcohols, most commonly ethanol, and less commonly propanol and butanol, are produced by the action of microorganisms and enzymes through the fermentation of sugars or starches (easiest), or cellulose (which is more difficult). Biobutanol (also called biogasoline) is often claimed to provide a direct replacement for gasoline, because it can be used directly in a gasoline engine.

Ethanol fuel is the most common biofuel worldwide, particularly in Brazil. Alcohol fuels are produced by fermentation of sugars derived from wheat, corn, sugar beets, sugar cane, molasses and any sugar or starch from which alcoholic beverages such as whiskey, can be made (such as potato and fruit waste, etc.). The ethanol production methods used are enzyme digestion (to release sugars from stored starches), fermentation of the sugars, distillation and drying. The distillation process requires significant energy input for heat (sometimes unsustainable natural gas fossil fuel, but cellulosic biomass such as bagasse, the waste left after sugar cane is pressed to extract its juice, is the most common fuel in Brazil, while pellets, wood chips and also waste heat are more common in Europe) Waste steam fuels ethanol factory – where waste heat from the factories also is used in the district heating grid.

Ethanol can be used in petrol engines as a replacement for gasoline; it can be mixed with gasoline to any percentage. Most existing car petrol engines can run on blends of up to 15% bioethanol with petroleum/gasoline. Ethanol has a smaller energy density than that of gasoline; this means it takes more fuel (volume and mass) to produce the same amount of work. An advantage of ethanol (CH_3CH_2OH) is that it has a higher octane rating than ethanol-free gasoline available at roadside gas stations, which allows an increase of an engine's compression ratio for increased thermal efficiency. In high-altitude (thin air) locations, some states mandate a mix of gasoline and ethanol as a winter oxidizer to reduce atmospheric pollution emissions.

Ethanol is also used to fuel bioethanol fireplaces. As they do not require a chimney and are "flueless", bioethanol fires are extremely useful for newly built homes and apartments without a flue. The downsides to these fireplaces is that their heat output is slightly less than electric heat or gas fires, and precautions must be taken to avoid carbon monoxide poisoning.

Corn-to-ethanol and other food stocks has led to the development of cellulosic ethanol. According to a joint research agenda conducted through the US Department of Energy, the fossil energy ratios (FER) for cellulosic ethanol, corn ethanol, and gasoline are 10.3, 1.36, and 0.81, respectively.

Ethanol has roughly one-third lower energy content per unit of volume compared to gasoline. This is partly counteracted by the better efficiency when using ethanol (in a long-term test of more than 2.1 million km, the BEST project found FFV vehicles to be 1–26% more energy efficient than petrol cars, but the volumetric consumption increases by approximately 30%, so more fuel stops are required).

With current subsidies, ethanol fuel is slightly cheaper per distance traveled in the United States.

Other Bioalcohols

Methanol is currently produced from natural gas, a non-renewable fossil fuel. In the future it is hoped to be produced from biomass as biomethanol. This is technically feasible, but the production is currently being postponed for concerns that the economic viability is still pending. The methanol economy is an alternative to the hydrogen economy, compared to today's hydrogen production from natural gas.

Butanol(C_4H_9OH) is formed by ABE fermentation (acetone, butanol, ethanol) and experimental modifications of the process show potentially high net energy gains with butanol as the only liquid product. Butanol will produce more energy and allegedly can be burned "straight" in existing gasoline engines (without modification to the engine or car), and is less corrosive and less water-soluble than ethanol, and could be distributed via existing infrastructures. DuPont and BP are working together to help develop butanol. *Escherichia coli* strains have also been successfully engineered to produce butanol by modifying their amino acid metabolism. One drawback to butanol production in E. coli remains the high cost of nutrient rich media, however, recent work has demonstrated E. coli can produce butanol with minimal nutritional supplementation.

Biodiesel

Biodiesel is the most common biofuel in Europe. It is produced from oils or fats using transesterification and is a liquid similar in composition to fossil/mineral diesel. Chemically, it consists mostly of fatty acid methyl (or ethyl) esters (FAMEs). Feedstocks for biodiesel include animal fats, vegetable oils, soy, rapeseed, jatropha, mahua, mustard, flax, sunflower, palm oil, hemp, field pennycress, *Pongamia pinnata* and algae. Pure biodiesel (B100, also known as "neat" biodiesel) currently reduces emissions with up to 60% compared to diesel Second generation B100.

Targray Biofuels Division railcar transporting Biodiesel.

Biodiesel can be used in any diesel engine when mixed with mineral diesel. It can also be used in its pure form (B100) in diesel engines, but some maintenance and performance problems may then occur during wintertime utilization, since the fuel becomes somewhat more viscous at lower temperatures, depending on the feedstock used.

In some countries, manufacturers cover their diesel engines under warranty for B100 use, although Volkswagen of Germany, for example, asks drivers to check by telephone with the VW environmental services department before switching to B100. In most cases, biodiesel is compatible with diesel engines from 1994 onwards, which use 'Viton' (by DuPont) synthetic rubber in their mechanical fuel injection systems. Note however, that no vehicles are certified for using pure biodiesel before 2014, as there was no emission control protocol available for biodiesel before this date.

Electronically controlled 'common rail' and 'unit injector' type systems from the late 1990s onwards may only use biodiesel blended with conventional diesel fuel. These engines have finely metered and atomized multiple-stage injection systems that are very sensitive to the viscosity of the fuel. Many current-generation diesel engines are made so that they can run on B100 without altering the engine itself, although this depends on the fuel rail design. Since biodiesel is an effective solvent and cleans residues deposited by mineral diesel, engine filters may need to be replaced more often, as the biofuel dissolves old deposits in the fuel tank and pipes. It also effectively cleans the engine combustion chamber of carbon deposits, helping to maintain efficiency. In many European countries, a 5% biodiesel blend is widely used and is available at thousands of gas stations. Biodiesel is also an oxygenated fuel, meaning it contains a reduced amount of carbon and higher hydrogen and oxygen content than fossil diesel. This improves the combustion of biodiesel and reduces the particulate emissions from unburnt carbon. However, using pure biodiesel may increase NO_x-emissions

Biodiesel is also safe to handle and transport because it is non-toxic and biodegradable, and has a high flash point of about 300 °F (148 °C) compared to petroleum diesel fuel, which has a flash point of 125 °F (52 °C).

In the US, more than 80% of commercial trucks and city buses run on diesel. The emerging US biodiesel market is estimated to have grown 200% from 2004 to 2005. "By the end of 2006 biodiesel production was estimated to increase fourfold to more than" 1 billion US gallons (3,800,000 m³).

In France, biodiesel is incorporated at a rate of 8% in the fuel used by all French diesel vehicles. Avril Group produces under the brand Diester, a fifth of 11 million tons of biodiesel consumed annually by the European Union. It is the leading European producer of biodiesel.

Green Diesel

Green diesel is produced through hydrocracking biological oil feedstocks, such as vegetable oils and animal fats. Hydrocracking is a refinery method that uses elevated temperatures and pressure in the presence of a catalyst to break down larger molecules, such as those found in vegetable oils, into shorter hydrocarbon chains used in diesel engines. It may also be called renewable diesel, hydrotreated vegetable oil or hydrogen-derived renewable diesel. Unlike biodiesel, green diesel has exactly the same chemical properties as petroleum-based diesel. It does not require new engines, pipelines or infrastructure to distribute and use, but has not been produced at a cost that is competitive with petroleum. Gasoline versions are also being developed. Green diesel is being developed in Louisiana and Singapore by ConocoPhillips, Neste Oil, Valero, Dynamic Fuels, and Honeywell UOP as well as Preem in Gothenburg, Sweden, creating what is known as Evolution Diesel.

Straight Vegetable Oil

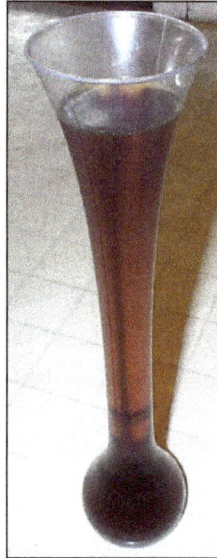

Filtered waste vegetable oil.

This truck is one of 15 based at Walmart's Buckeye, Arizona distribution
center that was converted to run on a biofuel made from reclaimed cooking grease
produced during food preparation at Walmart stores.

Straight unmodified edible vegetable oil is generally not used as fuel, but lower-quality oil has been used for this purpose. Used vegetable oil is increasingly being processed into biodiesel, or (more rarely) cleaned of water and particulates and then used as a fuel.

As with 100% biodiesel (B100), to ensure the fuel injectors atomize the vegetable oil in the correct pattern for efficient combustion, vegetable oil fuel must be heated to reduce its viscosity to that of diesel, either by electric coils or heat exchangers. This is easier in warm or temperate climates. MAN B&W Diesel, Wärtsilä, and Deutz AG, as well as a number of smaller companies, such as Elsbett, offer engines that are compatible with straight vegetable oil, without the need for after-market modifications.

Vegetable oil can also be used in many older diesel engines that do not use common rail or unit injection electronic diesel injection systems. Due to the design of the combustion chambers in indirect injection engines, these are the best engines for use with vegetable oil. This system allows the relatively larger oil molecules more time to burn. Some older engines, especially Mercedes, are driven experimentally by enthusiasts without any conversion, a handful of drivers have experienced limited success with earlier pre-"Pumpe Duse" VW TDI engines and other similar engines with direct injection. Several companies, such as Elsbett or Wolf, have developed professional conversion kits and successfully installed hundreds of them over the last decades.

Oils and fats can be hydrogenated to give a diesel substitute. The resulting product is a straight-chain hydrocarbon with a high cetane number, low in aromatics and sulfur and does not contain oxygen. Hydrogenated oils can be blended with diesel in all proportions. They have several advantages over biodiesel, including good performance at low temperatures, no storage stability problems and no susceptibility to microbial attack.

Bioethers

Bioethers (also referred to as fuel ethers or oxygenated fuels) are cost-effective compounds that act as octane rating enhancers."Bioethers are produced by the reaction of reactive iso-olefins, such as iso-butylene, with bioethanol." Bioethers are created from wheat or sugar beets. They also enhance engine performance, while significantly reducing engine wear and toxic exhaust emissions. Although bioethers are likely to replace petroethers in the UK, it is highly unlikely they will become a fuel in and of itself due to the low energy density. By greatly reducing the amount of ground-level ozone emissions, they contribute to air quality.

When it comes to transportation fuel there are six ether additives: dimethyl ether (DME), diethyl ether (DEE), methyl *tert*-butyl ether (MTBE), ethyl *tert*-butyl ether (ETBE), *tert*-amyl methyl ether (TAME), and *tert*-amyl ethyl ether (TAEE).

The European Fuel Oxygenates Association (EFOA) identifies methyl *tert*-butyl ether (MTBE) and ethyl *tert*-butyl ether (ETBE) as the most commonly used ethers in fuel to replace lead. Ethers were introduced in Europe in the 1970s to replace the highly toxic compound. Although Europeans still use bioether additives, the US no longer has an oxygenate requirement therefore bioethers are no longer used as the main fuel additive.

By Compatibility with Existing Infrastructure

So-called "drop-in" biofuels can be defined as "liquid bio-hydrocarbons that are functionally equivalent to petroleum fuels and are fully compatible with existing petroleum infrastructure". Drop-in biofuels require no (engine) modification of the vehicle.

Some examples of drop-in biofuels include biobutanol, biodiesel, synthetic paraffinic kerosine, and other synthetic fuels.

"The Potential and Challenges of Drop-in Biofuels", there are several ways to produce drop-in biofuels that are functionally equivalent to petroleum-derived transportation fuel blendstocks.

- Oleochemical processes, such as the hydroprocessing of lipid feedstocks obtained from oilseed crops, algae or tallow;

- Thermochemical processes, such as the thermochemical conversion of biomass to fluid intermediates (gas or oil) followed by catalytic upgrading and hydroprocessing to hydrocarbon fuels;

- Biochemical processes, such as the biological conversion of biomass (sugars, starches or lignocellulose-derived feedstocks) to longer chain alcohols and hydrocarbons.

A fourth category is also briefly described that includes "hybrid" thermochemical/biochemical technologies such as fermentation of synthesis gas and catalytic reforming of sugars/carbohydrates.

The report concludes by stating:

Tremendous entrepreneurial activity to develop and commercialize drop-in biofuels from aquatic and terrestrial feedstocks has taken place over the past several years. However, despite these efforts, drop-in biofuels represent only a small percentage (around 2%) of global biofuel markets. Due to the increased processing and resource requirements (e.g. hydrogen and catalysts) needed to make drop-in biofuels as compared to conventional biofuels, large scale production of cost-competitive drop-in biofuels is not expected to occur in the near to midterm. Rather, dedicated policies to promote development and commercialization of these fuels will be needed before they become significant contributors to global biofuels production. Currently, no policies (e.g. tax breaks, subsidies etc.) differentiate new, more fungible and infrastructure ready drop-in type biofuels from less infrastructure compatible oxygenated biofuels. Thus, while tremendous technical progress has been made in developing and improving the various routes to drop-in fuels, supportive policies directed specifically towards the further development of drop-in biofuels are likely to be needed to ensure their future commercial success.

Air Pollution

Biofuels are similar to fossil fuels in that biofuels contribute to air pollution. Burning produces carbon dioxide, airborne carbon particulates, carbon monoxide and nitrous oxides. The WHO estimates 3.7 million premature deaths worldwide in 2012 due to air pollution. Brazil burns significant amounts of ethanol biofuel. Gas chromatograph studies were performed of ambient air in São Paulo, Brazil, and compared to Osaka, Japan, which does not burn ethanol fuel. Atmospheric Formaldehyde was 160% higher in Brazil, and Acetaldehyde was 260% higher.

The Environmental Protection Agency acknowledged in April 2007 that the increased use of bioethanol will lead to worse air quality. The total emissions of air pollutants such as nitrogen oxides will rise due the growing use of bioethanol. There is an increase in carbon dioxide from the burning of fossil fuels to produce the biofuels as well as nitrous oxide from the soil, which has most likely been treated with nitrogen fertilizer. Nitrous oxide is known to have a greater impact on the atmosphere in relation to global warming, as it is also an ozone destroyer.

Advantages of Biofuels

Cost Benefit

As of now, biofuels cost the same in the market as gasoline does. However, the overall cost benefit of using them is much higher. They are cleaner fuels, which means they produce fewer emissions on burning. Biofuels are adaptable to current engine designs and perform very well in most conditions. This keeps the engine running for longer, requires less maintenance and brings down overall pollution check costs. With the increased demand of biofuels, they have a potential of becoming cheaper in future as well. So, the use of biofuels will be less of a drain on the wallet.

Easy to Source

Gasoline is refined from crude oil, which happens to be a non-renewable resource. Although current reservoirs of gas will sustain for many years, they will end sometime in near future. Biofuels are made from many different sources such as manure, waste from crops and plants grown specifically for the fuel.

Renewable

Most of the fossil fuels will expire and end up in smoke one day. Since most of the sources like manure, corn, switchgrass, soyabeans, waste from crops and plants are renewable and are not likely to run out any time soon, making the use of biofuels efficient in nature. These crops can be replanted again and again.

Reduce Greenhouse Gases

Fossil fuels, when burnt, produce large amount of greenhouse gases i.e. carbon dioxide in the atmosphere. These greenhouse gases trap sunlight and cause planet to warm. The burning of coal and oil increases the temperature and causes global warming. To reduce the impact of greenhouse gases, people around the world are using biofuels. Studies suggests that biofuels reduces greenhouse gases up to 65 percent.

Economic Security

Not every country has large reserves of crude oil. For them, having to import the oil puts a huge dent in the economy. If more people start shifting towards biofuels, a country can reduce its dependance on fossil fuels. More jobs will be created with a growing biofuel industry, which will keep our economy secure.

Reduce Dependance on Foreign Oil

While locally grown crops has reduce the nation's dependance on fossil fuels, many experts believe that it will take a long time to solve our energy needs. As prices of crude oil is touching sky high, we need some more alternative energy solutions to reduce our dependance on fossil fuels.

Lower Levels of Pollution

Since biofuels can be made from renewable resources, they cause less pollution to the planet. However, that is not the only reason why the use of biofuels is being encouraged. They release lower levels of carbon dioxide and other emissions when burnt. Although the production of biofuels creates carbon dioxide as a byproduct, it is frequently used to grow the plants that will be converted into the fuel. This allows it to become something close to a self sustaining system.

Disadvantages of Biofuels

High Cost of Production

Even with all the benefits associated with biofuels, they are quite expensive to produce in the current market. As of now, the interest and capital investment being put into biofuel production is fairly low but it can match demand. If the demand increases, then increasing the supply will be a long term operation, which will be quite expensive. Such a disadvantage is still preventing the use of biofuels from becoming more popular.

Monoculture

Monoculture refers to practice of producing same crops year after year, rather than producing various crops through a farmer's fields over time. While, this might be economically attractive for farmers but growing same crop every year may deprive the soil of nutrients that are put back into the soil through crop rotation.

Use of Fertilizers

Biofuels are produced from crops and these crops need fertilizers to grow better. The downside of using fertilizers is that they can have harmful effects on surrounding environment and may cause water pollution. Fertilizers contain nitrogen and phosphorus. They can be washed away from soil to nearby lake, river or pond.

Shortage of Food

Biofuels are extracted from plants and crops that have high levels of sugar in them. However, most of these crops are also used as food crops. Even though waste material from plants can be used as raw material, the requirement for such food crops will still exist. It will take up agricultural space from other crops, which can create a number of problems. Even if it does not cause an acute shortage of food, it will definitely put pressure on the current growth of crops. One major worry being faced by people is that the growing use of biofuels may just mean a rise in food prices as well.

Industrial Pollution

The carbon footprint of biofuels is less than the traditional forms of fuel when burnt. However, the process with which they are produced makes up for that. Production is largely dependent on lots of water and oil. Large scale industries meant for churning out biofuel are known to emit large amounts of emissions and cause small scale water pollution as well. Unless more efficient means of production are put into place, the overall carbon emission does not get a very big dent in it.

Water Use

Large quantities of water are required to irrigate the biofuel crops and it may impose strain on local and regional water resources, if not managed wisely. In order to produce corn based ethanol to meet local demand for biofuels, massive quantities of water are used that could put unsustainable pressure on local water resources

Future Rise in Price

Current technology being employed for the production of biofuels is not as efficient as it should be. Scientists are engaged in developing better means by which we can extract this fuel. However, the cost of research and future installation means that the price of biofuels will see a significant spike. As of now, the prices are comparable with gasoline and are still feasible. Constantly rising prices may make the use of biofuels as harsh on the economy as the rising gas prices are doing right now.

BIOMASS AND THE ENVIRONMENT

Biomass and biofuels made from biomass are alternative energy sources to fossil fuels—coal, petroleum, and natural gas. Burning either fossil fuels or biomass releases carbon dioxide (CO_2), a greenhouse gas. However, the plants that are the source of biomass capture a nearly equivalent amount of CO_2 through photosynthesis while they are growing, which can make biomass a carbon-neutral energy source.

Burning Wood

Using wood, wood pellets, and charcoal for heating and cooking can replace fossil fuels and may result in lower CO_2 emissions overall. Wood can be harvested from forests, from woodlots that have to be thinned, or from urban trees that fall down or have to be cut down.

Wood smoke contains harmful pollutants such as carbon monoxide and particulate matter. Modern wood-burning stoves, pellet stoves, and fireplace inserts can reduce the amount of particulates from burning wood. Wood and charcoal are major cooking and heating fuels in poor countries, but if people harvest the wood faster than trees can grow, it causes deforestation. Planting fast-growing trees for fuel and using fuel-efficient cooking stoves can help slow deforestation and improve the environment.

Switchgrass growing on a test plot for biomass production.

Burning Municipal Solid Waste or Wood Waste

Burning municipal solid waste (MSW, or garbage) in waste-to-energy plants could result in less waste buried in landfills. On the other hand, burning garbage produces air pollution and releases the chemicals and substances in the waste into the air. Some of these chemicals can be hazardous to people and the environment if they are not properly controlled.

The Environmental Protection Agency (EPA) applies strict environmental rules to waste-to-energy plants, which require waste-to-energy plants to use air pollution control devices such as scrubbers, fabric filters, and electrostatic precipitators to capture air pollutants.

Scrubbers clean emissions from waste-to-energy facilities by spraying a liquid into the combustion gases to neutralize the acids present in the stream of emissions. Fabric filters and electrostatic precipitators also remove particles from the combustion gases. The particles—called fly ash—are then mixed with the ash that is removed from the bottom of the waste-to-energy furnace.

A waste-to-energy furnace burns at high temperatures (1,800 °F to 2,000 °F), which breaks down the chemicals in MSW into simpler, less harmful compounds.

Disposing Ash from Waste-to-Energy Plants

Ash from waste-to-energy plants can contain high concentrations of various metals that were present in the original waste. Textile dyes, printing inks, and ceramics, for example, may contain lead and cadmium.

Separating waste before burning can solve part of the problem. Because batteries are the largest source of lead and cadmium in municipal waste, they should not be included in regular trash. Florescent light bulbs should also not be put in regular trash because they contain small amounts of mercury.

The EPA tests ash from waste-to-energy plants to make sure that it is not hazardous. The test looks for chemicals and metals that could contaminate ground water. Some MSW landfills use ash that is considered safe as a cover layer for their landfills, and some MSW ash is used to make concrete blocks and bricks.

Collecting Landfill Gas or Biogas

Biogas forms as a result of biological processes in sewage treatment plants, waste landfills, and livestock manure management systems. Biogas is composed mainly of methane (a greenhouse gas) and CO_2. Many facilities that produce biogas capture it and burn the methane for heat or to generate electricity. This electricity is considered renewable and, in many states, contributes to meeting state renewable portfolio standards (RPS). This electricity may replace electricity generation from fossil fuels and can result in a net reduction in CO_2 emissions. Burning methane produces CO_2, but because methane is a stronger greenhouse gas than CO_2, the overall greenhouse effect is lower.

Liquid Biofuels: Ethanol and Biodiesel

Biofuels are transportation fuels such as ethanol and biodiesel. The federal government promotes biofuels as transportation fuels to help reduce oil imports and CO_2 emissions. In 2007, the government set a target to use 36 billion gallons of biofuels by 2022. As a result, nearly all gasoline now sold in the United States contains some ethanol.

Biofuels may be carbon-neutral because the plants that are used to make biofuels (such as corn and sugarcane for ethanol and soy beans and oil palm trees for biodiesel) absorb CO_2 as they grow and may offset the CO_2 emissions when biofuels are produced and burned.

Growing plants for biofuels is controversial because the land, fertilizers, and energy for growing biofuel crops could be used to grow food crops instead. In some parts of the world, large areas of natural vegetation and forests have been cut down to grow sugar cane for ethanol and soybeans and oil palm trees for biodiesel. The government supports efforts to develop alternative sources of biomass that do not compete with food crops and that use less fertilizer and pesticides than corn and sugar cane. The government also supports methods to produce ethanol that require less energy than conventional fermentation. Ethanol can also be made from waste paper, and biodiesel can be made from waste grease and oils and even algae.

Ethanol and gasoline-ethanol blends burn cleaner and have higher octane ratings than pure gasoline, but they have higher evaporative emissions from fuel tanks and dispensing equipment. These evaporative emissions contribute to the formation of harmful, ground-level ozone and smog. Gasoline requires extra processing to reduce evaporative emissions before it is blended with ethanol. Biodiesel combustion produces fewer sulfur oxides, less particulate matter, less carbon monoxide, and fewer unburned and other hydrocarbons, but it does produce more nitrogen oxide than petroleum diesel.

BIOENERGY TECHNOLOGIES

A wide range of technologies are available for realizing the energy potential of biomass wastes, ranging from very simple systems for disposing of dry waste to more complex technologies capable of dealing with large amounts of industrial waste. Conversion routes for biomass wastes are generally thermo-chemical or bio-chemical, but may also include chemical and physical.

Thermal Technologies

The three principal methods of thermo-chemical conversion corresponding to each of these energy carriers are combustion in excess air, gasification in reduced air, and pyrolysis in the absence of air. Direct combustion is the best established and most commonly used technology for converting wastes to heat. During combustion, biomass is burnt in excess air to produce heat. The first stage of combustion involves the evolution of combustible vapours from wastes, which burn as flames. Steam is expanded through a conventional turbo-alternator to produce electricity. The residual material, in the form of charcoal, is burnt in a forced air supply to give more heat.

Co-firing or co-combustion of biomass wastes with coal and other fossil fuels can provide a short-term, low-risk, low-cost option for producing renewable energy while simultaneously reducing the use of fossil fuels. Co-firing involves utilizing existing power generating plants that are fired with fossil fuel (generally coal), and displacing a small proportion of the fossil fuel with renewable biomass fuels. Co-firing has the major advantage of avoiding the construction of new, dedicated, waste-to-energy power plant. An existing power station is modified to accept the waste resource and utilize it to produce a minor proportion of its electricity.

Gasification systems operate by heating biomass wastes in an environment where the solid waste breaks down to form a flammable gas. The gasification of biomass takes place in a restricted supply of air or oxygen at temperatures up to 1200–1300°C. The gas produced—synthesis gas, or syngas—can be cleaned, filtered, and then burned in a gas turbine in simple or combined-cycle mode, comparable to LFG or biogas produced from an anaerobic digester. The final fuel gas consists principally of carbon monoxide, hydrogen and methane with small amounts of higher hydrocarbons. This fuel gas may be burnt to generate heat; alternatively it may be processed and then used as fuel for gas-fired engines or gas turbines to drive generators. In smaller systems, the syngas can be fired in reciprocating engines, micro-turbines, Stirling engines, or fuel cells.

Pyrolysis is thermal decomposition occurring in the absence of oxygen. During the pyrolysis process, biomass waste is heated either in the absence of air (i.e. indirectly), or by the partial combustion of some of the waste in a restricted air or oxygen supply. This results in the thermal decomposition of the waste to form a combination of a solid char, gas, and liquid bio-oil, which can be used as a liquid fuel or upgraded and further processed to value-added products.

Biochemical Technologies

Biochemical processes, like anaerobic digestion, can also produce clean energy in the form of biogas which can be converted to power and heat using a gas engine. Anaerobic digestion is a series of chemical reactions during which organic material is decomposed through the metabolic pathways of naturally occurring microorganisms in an oxygen depleted environment. In addition, wastes can also yield liquid fuels, such as cellulosic ethanol and biodiesel, which can be used to replace petroleum-based fuels.

Anaerobic digestion is the natural biological process which stabilizes organic waste in the absence of air and transforms it into biogas and biofertilizer. Almost any organic material can be processed with anaerobic digestion. This includes biodegradable waste materials such as municipal solid waste, animal manure, poultry litter, food wastes, sewage and industrial wastes. An anaerobic digestion plant produces two outputs, biogas and digestate, both can be further processed or utilized to produce secondary outputs. Biogas can be used for producing electricity and heat, as a natural gas substitute and also a transportation fuel. Digestate can be further processed to produce liquor and a fibrous material. The fiber, which can be processed into compost, is a bulky material with low levels of nutrients and can be used as a soil conditioner or a low level fertilizer.

A variety of fuels can be produced from biomass wastes including liquid fuels, such as ethanol, methanol, biodiesel, Fischer-Tropsch diesel, and gaseous fuels, such as hydrogen and methane. The resource base for biofuel production is composed of a wide variety of forestry and agricultural resources, industrial processing residues, and municipal solid and urban wood residues. The largest potential feedstock for ethanol is lignocellulosic biomass wastes, which includes materials such as agricultural residues (corn stover, crop straws and bagasse), herbaceous crops (alfalfa, switchgrass), short rotation woody crops, forestry residues, waste paper and other wastes (municipal and industrial). The three major steps involved in cellulosic ethanol production are pretreatment, enzymatic hydrolysis, and fermentation. Biomass is pretreated to improve the accessibility of enzymes. After pretreatment, biomass undergoes enzymatic hydrolysis for conversion of polysaccharides into monomer sugars, such as glucose and xylose. Subsequently, sugars are fermented to ethanol by the use of different microorganisms. Bioethanol production from these feedstocks could be an attractive alternative for disposal of these residues. Importantly, lignocellulosic feedstocks do not interfere with food security.

References

- Uses-of-biomass-energy: mainrenewableenergy.Com, retrieved 18 april, 2019

- "Technology: high yield carbon recycling". Greenfuel technologies corporation. Archived from the original on 21 august 2007. Retrieved 9 july 2008

- Advantages-and-disadvantages-of-biofuels: conserve-energy-future.Com, retrieved 21 july, 2019

- The royal society (january 2008). Sustainable biofuels: prospects and challenges, isbn 978-0-85403-662-2, p. 61

- Biomass-conversion, tag: bioenergyconsult.Com, retrieved 27 may, 2019

6

Environmental Impacts of Renewable Energy

The impact of renewable energy upon the environment depends on numerous factors such as the technologies used and geographic locations. The environmental impacts of various renewable sources of energy such as solar energy, wind power, hydroelectric power and geothermal power have been thoroughly discussed in this chapter.

IMPACT OF SOLAR ENERGY ON THE ENVIRONMENT

The sun is a huge source of energy which has only recently been tapped into. It provides immense resources which can generate clean, non-polluting and sustainable electricity, thus resulting in no global warming emissions. In recent years, it was discovered that solar energy can be collected and stored, to be used on a global scale with the purpose of eventually replacing the conventional sources of energy. As the world is turning its focus to cleaner power, solar energy has seen a significant rise in importance.

With the recent announcement of the VAT increase on solar panels in the UK, we urge you to invest now to avoid the extra costs.

Solar energy systems offer significant environmental benefits in comparison to the conventional energy sources, thus they greatly contribute to the sustainable development of human activities. At times however, the wide scale deployment of such systems has to face potential negative environmental implications. These possible problems may be a strong barrier for further advancement of these systems in some consumers.

The potential environmental impacts associated with solar power can be classified according to numerous categories, some of which are land use impacts, ecological impacts, impacts to water, air and soil, and other impacts such as socioeconomic ones, and can vary greatly depending on the technology, which includes two broad categories:

- Photovoltaic (PV) solar panels.
- Concentrating solar thermal plants (CSP).

Environmental Impacts of Solar Energy

Land use and Ecological Impacts

In the point of generating electricity at a utility-scale, solar energy facilities necessitate large areas for collection of energy. Due to this, the facilities may interfere with existing land uses and can impact the use of areas such as wilderness or recreational management areas.

As energy systems may impact land through materials exploration, extraction, manufacturing and disposal, energy footprints can become incrementally high. Thus, some of the lands may be utilised for energy in such a way that returning to a pre-disturbed state necessitates significant energy input or time, or both, whereas other uses are so dramatic that incurred changes are irreversible.

Impacts to Soil, Water and Air Resources

The construction of solar facilities on vast areas of land imposes clearing and grading, resulting in soil compaction, alteration of drainage channels and increased erosion. Central tower systems require consuming water for cooling, which is a concern in arid settings, as an increase in water demand may strain available water resources as well as chemical spills from the facilities which may result in the contamination of groundwater or the ground surface.

As with the development of any large-scale industrial facility, the construction of solar energy power plants can pose hazards to air quality. Such threats include the release of soil-carried pathogens and results in an increase in air particulate matter which has the effect of contaminating water reservoirs.

Heavy Metals

Some have argued that the latest technologies introduced on the market, namely thin-film panels, are manufactured using dangerous heavy metals, such as Cadium

Telluride. While it is true that solar panel manufacturing uses these dangerous material, coal and oil also contain the same substsances, which are released with combustion.

Moreover, coal power plants emit much more of these toxic substances, polluting up to 300 times more than solar panel manufacturers.

Other Impacts

Besides the aforementioned environmental impacts, solar energy facilities also may have other impacts, such as influencing the socio-economic state of an area. Construction and operation of utility-scale solar energy facilities in an area would produce direct and indirect economic impacts.

- The direct impacts would occur as a result of expenses on wages and salaries as well as the attaining of goods and services which are required for project construction and operation.

- Indirect impacts would occur in the form of project wages and salaries procurement expenditures, which create additional employment, income, and tax revenues. Facility construction and operation would require in-migration of workers, affecting housing, public services, and local government employment.

Recycling Solar Panels

Currently the recycling of solar panels faces a big issue, specifically, there aren't enough locations to recycle old solar panels, and there aren't enough non-operational solar panels to make recycling them economically attractive. Recycling of solar panels is particularly important because the materials used to make the panels are rare or precious metals, all of them being composed of silver, tellurium, or indium. Due to the limitability of recycling the panels, those recoverable metals may be going to waste which may result in resource scarcity issues in the future.

Looking at silicon for example, one resource that is needed to make the majority of present day photovoltaic cells and which there is currently an abundance of, however a silicon-based solar cell requires a lot of energy input in its manufacturing process, the source of that energy, which is often coal, determining how large the cell's carbon footprint is.

The lack of awareness regarding the manufacturing process of solar panels and to the issue of recycling these, as well as the absence of much external pressure are the causes of the insufficiency in driving significant change in the recycling of the materials used in solar panel manufacturing, a business that, from a power-generation standpoint, already has great environmental credibility.

ENVIRONMENTAL IMPACTS OF WIND POWER

The environmental impact of wind power when compared to that of fossil fuel power, is relatively minor. Compared with other low carbon power sources, wind turbines have some of the lowest global warming potential per unit of electrical energy generated. According to the IPCC, in assessments of the life-cycle global warming potential of energy sources, wind turbines have a median value of between 15 and 11 (gCO_2eq/kWh) depending on whether off- or onshore turbines are being assessed.

Onshore wind farms can have a significant impact on the landscape, as typically they need to be spread over more land than other power stations and need to be built in wild and rural areas, which can lead to "industrialization of the countryside" and habitat loss. Conflicts arise especially in scenic and culturally-important landscapes. Siting restrictions (such as setbacks) may be implemented to limit the impact. Land between the turbines and access roads can still be used for farming and grazing.

Habitat loss and fragmentation are the greatest impact of wind farms on wildlife. Wind turbines, like many other human activities and buildings, also increase bat and bird deaths. A summary of the existing field studies compiled in 2010 from the National Wind Coordinating Collaborative identified fewer than 14 and typically less than 4 bird deaths per installed megawatt per year, but a wider variation in the number of bat deaths. Like other investigations, it concluded that some species (e.g. migrating bats and songbirds) are known to be harmed more than others and that factors such as turbine siting can be important. However, many details as well as the overall impact from the growing number of turbines remain unclear.

Wind turbines generate some noise. At a residential distance of 300 metres (980 ft) this may be around 45 dB. At 1.5 km (1 mi) distance most wind turbines become inaudible. Loud or persistent noise increases stress, and stress causes diseases. Wind turbines do not affect human health from noise when properly placed. However, when improperly sited, data from the monitoring of two groups of growing geese revealed substantially lower body weights and higher concentrations of a stress hormone in the blood of the first group of geese who were situated 50 meters away compared to a second group which was at a distance of 500 meters from the turbine.

Basic Operational Considerations

Net Energy Gain

The energy return on investment (EROI) for wind energy is equal to the cumulative electricity generated divided by the cumulative primary energy required to build and maintain a turbine. According to a meta study, in which all existing studies from 1977 to 2007 were reviewed, the EROI for wind ranges from 5 to 35, with the most common turbines in the range of 2 MW nameplate capacity-rotor diameters of 66 meters, on average the EROI is 16. EROI is strongly proportional to turbine size, and larger late-generation turbines average at the high end of this range, are by one study, approximately 35.

Wind turbine manufacturer Vestas claims that initial energy "pay back" is within about 7–9 months of operation for a 1.65-2.0MW wind turbine under low wind conditions, whereas Siemens Wind Power calculates 5–10 months depending on circumstances.

Pollution and Effects on the Grid

Pollution Costs

Wind power consumes no water for continuing operation, and has near negligible emissions directly related to its electricity production. Wind turbines when isolated from the electric grid produce negligible amounts of carbon dioxide, carbon monoxide, sulfur dioxide, nitrogen dioxide, mercury and radioactive waste when in operation, unlike fossil fuel sources and nuclear energy station fuel production, respectively.

With the construction phase largely to blame, wind turbines emit slightly more particulate matter (PM), a form of air pollution, at an "exception" rate higher per unit of energy generated(kWh) than a fossil gas electricity station("NGCC"), and also emit more heavy metals and PM than nuclear stations, per unit of energy generated. As far as total pollution costs in economic terms, in a comprehensive 2006 European study, alpine Hydropower was found to exhibit the lowest external pollution, or externality, costs of all electricity generating systems, below 0.05 c€/kWh. Wind power externality costs were found to be 0.09 - 0.12c€/kW, while nuclear energy had a 0.19 c€/kWh value and fossil fuels generated 1.6 - 5.8 c€/kWh of downstream costs. With the exception of the latter fossil fuels, these are negligible costs in comparison to the cost of electricity production, which is approximately 10 c€/kWh in European countries.

Findings when Connected to the Grid

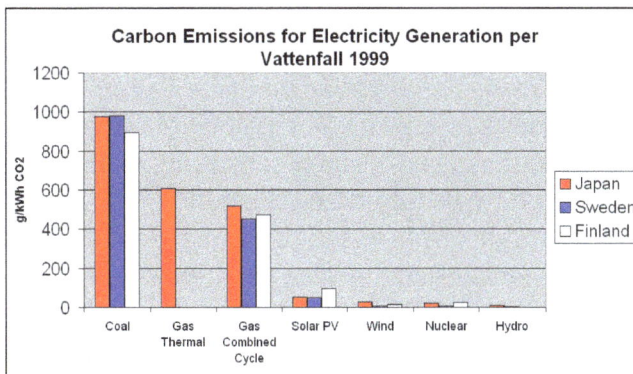

The Vattenfall utility company study found Hydroelectric, nuclear stations and wind turbines to have far less greenhouse emissions than other sources represented.

A typical study of a wind farm's Life cycle assessment, when not connected to the electric grid, usually results in similar findings as the following 2006 analysis of 3 installations in the US Midwest, where the carbon dioxide (CO_2) emissions of wind power ranged from 14 to 33 tonnes (15 to 36 short tons) per GWh (14–33 gCO_2/kWh) of energy produced, with most of the CO_2 emission intensity coming from producing the concrete for wind-turbine foundations. By combining similar data from numerous individual studies in a meta-analysis, the median global warming potential for wind power was found to be 11–12 g CO_2/kWh and unlikely to change significantly.

However these relatively low pollution values begin to increase as greater and greater wind energy is added to the grid, or wind power 'electric grid penetration' levels are reached. Due to the effects of attempting to balance out the energy demands on the grid, from Intermittent power sources e.g. wind power (sources which have low capacity factors due to the weather), this either requires the construction of large energy storage projects, which have their own emission intensity which must be added to wind power's system-wide pollution effects, or it requires more frequent reliance on fossil fuels than the spinning reserve requirements necessary to back up more dependable sources. With the latter combination presently being the more common.

This higher dependence on back-up/Load following power plants to ensure a steady power grid output has the knock-on-effect of more frequent inefficient (in CO_2e g/kWh) throttling up and down of these other power sources in the grid to facilitate the intermittent power source's variable output. When one includes the total effect of intermittent sources on other power sources in the grid system, that is, including these inefficient start up emissions of backup power sources to cater for wind energy, into wind energy's total system wide life cycle, this results in a higher real-world wind energy emission intensity. Higher than the direct g/kWh value that is determined from looking at the power source in isolation and thus ignores all down-stream detrimental/inefficiency effects it has on the grid. This higher dependence on back-up/Load following power plants to ensure a steady power grid output forces fossil power plants to operate in less efficient states.

The thermal efficiency of fossil-based power plants is reduced when operated at fluctuating and suboptimal loads to supplement wind power, which may degrade, to a certain extent, the GHG(Greenhouse gas) benefits resulting from the addition of wind to the grid. A study conducted by Pehnt and colleagues (2008) reports that a moderate level of [grid] wind penetration (12%) would result in efficiency penalties of 3% to 8%, depending on the type of conventional power plant considered. Gross and colleagues (2006) report similar results, with efficiency penalties ranging from nearly 0% to 7% for up to 20% [of grid] wind penetration. Pehnt and colleagues (2008) conclude that the results of adding offshore wind power in Germany on the background power systems maintaining a level supply to the grid and providing enough reserve capacity amount to adding between 20 and 80 g CO_2-eq/kWh to the life cycle GHG emissions profile of wind power.

In comparison to other low carbon power sources Wind turbines, when assessed in isolation, have a median life cycle emission value of between 11 and 12 (gCO_2eq/kWh). The more dependable alpine Hydropower and nuclear stations have median total life cycle emission values of 24 and 12 g CO_2-eq/kWh respectively.

While an increase in emissions due to the practical issues of load balancing is an issue, Pehnt et al. still conclude that these 20 and 80 g CO_2-eq/kWh added penalties still result in wind being roughly ten times less polluting than fossil gas and coal which emit ~400 and 900 g CO_2-eq/kWh respectively.

As these losses occur due to cycling of fossil power plants, they may at some point become smaller when more than 20–30% of wind energy is added to the power grid, as fossil power plants are replaced, however this has yet to occur in practice.

Rare-Earth use

The production of permanent magnets used in some wind turbines makes use of neodymium. Primarily exported by China, pollution concerns associated with the extraction of this rare-earth element have prompted government action in recent years, and international research attempts to refine the extraction process. Research is underway on turbine and generator designs which reduce the need for neodymium, or eliminate the use of rare-earth metals altogether. Additionally, the large wind turbine manufacturer Enercon GmbH chose very early not to use permanent magnets for its direct drive turbines, in order to avoid responsibility for the adverse environmental impact of rare-earth mining.

Ecology

Land use

Wind farms are often built on land that has already been impacted by land clearing. The vegetation clearing and ground disturbance required for wind farms is minimal compared with coal mines and coal-fired power stations. If wind farms are decommissioned, the landscape can be returned to its previous condition.

A study by the US National Renewable Energy Laboratory of US wind farms built between 2000 and 2009 found that, on average, only 1.1 percent of the total wind farm area suffered surface disturbance, and only 0.43 percent was permanently disturbed by wind power installations. On average, there were 63 hectares (156 acres) of total wind farm area per MW of capacity, but only 0.27 hectares (0.67 acres) of permanently disturbed area per MW of wind power capacity.

In the UK many prime wind farm sites - locations with the best average wind speeds are in upland areas which are frequently covered by blanket bog. This type of habitat exists in areas of relatively high rainfall where large areas of land remain permanently sodden. Construction work may create a risk of disruption to peatland hydrology which could cause localised areas of peat within the area of a wind farm to dry out, disintegrate, and so release their stored carbon. At the same time, the warming climate which renewable energy schemes seek to mitigate could itself pose an existential threat to peatlands throughout the UK. A Scottish MEP campaigned for a moratorium on wind developments on peatlands saying that "Damaging the peat causes the release of more carbon dioxide than wind farms save". A 2014 report for the Northern Ireland Environment Agency noted that siting wind turbines on peatland could release considerable carbon dioxide from the peat, and also damage the peatland contributions to flood control and water quality: "The potential knock-on effects of using the peatland resource for wind turbines are considerable and it is arguable that the impacts on this facet of biodiversity will have the most noticeable and greatest financial implications for Northern Ireland."

Wind-energy advocates contend that less than 1% of the land is used for foundations and access roads, the other 99% can still be used for farming. A wind turbine needs about 200–400 m² for the foundation. A (small) 500-kW-turbine with an annual production of 1.4 GWh produces 11.7 MWh/m², which is comparable with coal-fired plants (about 15-20 MWh/m²), coal-mining not included. With increasing size of the wind turbine the relative size of the foundation decreases. Critics point out that on some locations in forests the clearing of trees around tower bases may be necessary for installation sites on mountain ridges, such as in the northeastern U.S. This usually takes the clearing of 5,000 m² per wind turbine.

Turbines are not generally installed in urban areas. Buildings interfere with wind, turbines must be sited a safe distance ("setback") from residences in case of failure, and the value of land is high. There are a few notable exceptions to this. The WindShare Ex-Place wind turbine was erected in December 2002, on the grounds of Exhibition Place, in Toronto, Ontario, Canada. It was the first wind turbine installed in a major North American urban city centre. Steel Winds also has a 20 MW urban project south of Buffalo, New York. Both of these projects are in urban locations, but benefit from being on uninhabited lake shore property.

Livestock

The land can still be used for farming and cattle grazing. Livestock are unaffected by the presence of wind farms. International experience shows that livestock will "graze right up to the base of wind turbines and often use them as rubbing posts or for shade".

In 2014, a first of its kind Veterinary study attempted to determine the effects of rearing livestock near a wind turbine, the study compared the health effects of a wind turbine on the development of two groups of growing geese, preliminary results found that geese raised within 50 meters of a wind turbine gained less weight and had a higher concentration of the stress hormone cortisol in their blood than geese at a distance of 500 meters.

Semi-domestic reindeer avoid the construction activity, but seem unaffected when the turbines are operating.

Impact on wildlife

Environmental assessments are routinely carried out for wind farm proposals, and potential impacts on the local environment (e.g. plants, animals, soils) are evaluated. Turbine locations and operations are often modified as part of the approval process to avoid or minimise impacts on threatened species and their habitats. Any unavoidable impacts can be offset with conservation improvements of similar ecosystems which are unaffected by the proposal.

A research agenda from a coalition of researchers from universities, industry, and government, supported by the Atkinson Center for a Sustainable Future, suggests modeling the spatiotemporal patterns of migratory and residential wildlife with respect to geographic features and weather, to provide a basis for science-based decisions about where to site new wind projects. More specifically, it suggests:

- Use existing data on migratory and other movements of wildlife to develop predictive models of risk.

- Use new and emerging technologies, including radar, acoustics, and thermal imaging, to fill gaps in knowledge of wildlife movements.

- Identify specific species or sets of species most at risk in areas of high potential wind resources.

Birds

The impact of wind energy on birds, which can fly into turbines directly, or indirectly have their habitats degraded by wind development, is complex. Projects such as the Black Law Wind Farm have received wide recognition for its contribution to environmental objectives, including praise from the Royal Society for the Protection of Birds, who describe the scheme as both improving the landscape of a derelict opencast mining site and also benefiting a range of wildlife in the area, with an extensive habitat management projects covering over 14 square kilometres.

The meta-analysis on avian mortality by Benjamin K. Sovacool led him to suggest that there were a number of deficiencies in other researchers' methodologies. Among them, he stated were a focus on bird deaths, but not on the reductions in bird births: for example, mining activities for fossil fuels and pollution from fossil fuel plants have led to significant toxic deposits and acid rain that have damaged or poisoned many nesting and feeding grounds, leading to reductions in births. The large cumulated footprint of wind turbines, which reduces the area available to wildlife or agriculture, is also missing from all studies including Sovacool's. Many of the studies also made no mention of avian deaths per unit of electricity produced, which excluded meaningful comparisons between different energy sources. More importantly, it concluded, the most visible impacts of a technology, as measured by media exposure, are not necessarily the most flagrant ones.

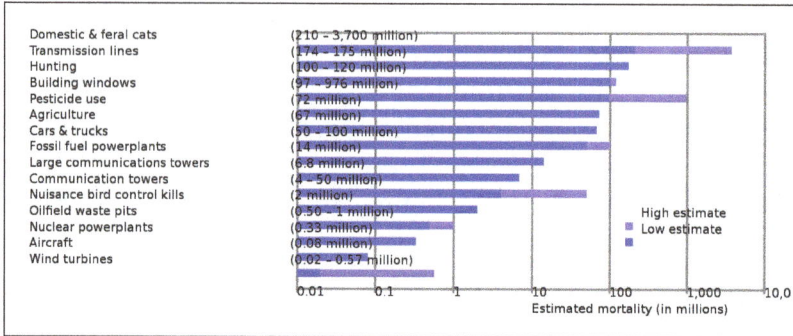

	Low estimate	High estimate
Domestic & feral cats	(210 – 3,700 million)	
Transmission lines	(174 – 175 million)	
Hunting	(100 – 120 million)	
Building windows	(97 – 976 million)	
Pesticide use	(72 million)	
Agriculture	(67 million)	
Cars & trucks	(50 – 100 million)	
Fossil fuel powerplants	(14 million)	
Large communications towers	(6.8 million)	
Communication towers	(4 – 50 million)	
Nuisance bird control kills	(2 million)	
Oilfield waste pits	(0.50 – 1 million)	
Nuclear powerplants	(0.33 million)	
Aircraft	(0.08 million)	
Wind turbines	(0.02 – 0.57 million)	

Estimated mortality (in millions)

The preliminary data, from the above table during 2013, 'Causes of avian mortality in the United States, annual', shown as a bar graph, inclusive of a high nuclear-fission bird mortality figure that the author later recognized was due to a major error on their part.

Sovacool estimated that in the United States wind turbines kill between 20,000 and 573,000 birds per year, and has stated he regards either figure as minimal compared to bird deaths from other causes. He uses the lower 20,000 figure in his study and table to arrive at a direct mortality rate per unit of energy generated figure of 0.269 per GWh for wind power. Fossil-fueled power plants, which wind turbines generally require to make up for their weather dependent intermittency, kill almost 20 times as many birds per gigawatt hour (GWh) of electricity according to Sovacool. Bird deaths due to other human activities and cats total between 797 million and 5.29 billion per year in the U.S. Additionally, while many studies concentrate on the analysis of bird deaths, few have been conducted on the reductions of bird births, which are the additional consequences of the various pollution sources that wind power partially mitigates.

Of the bird deaths Sovacool attributed to fossil-fuel power plants, 96 percent were due to the effects of climate change. While the study did not assess bat mortality due to various forms of energy, he considered it not unreasonable to assume a similar ratio of mortality. The Sovacool study has provoked controversy because of its treatment of data. In a series of replies, Sovacool acknowledged a number of large errors, particularly those that relate to his earlier "0.33 to 0.416" fatalities overestimate for the number of bird deaths per GWh of nuclear power, and cautioned that "the study already tells you the numbers are very rough estimates that need to be improved."

A 2013 meta-analysis by Smallwood identified a number of factors which result in serious under-reporting of bird and bat deaths by wind turbines. These include inefficient searches, inadequate search radius, and carcass removal by predators. To adjust the results of different studies, he applied correction factors from hundreds of carcass placement trials. His meta-analysis concluded that in 2012 in the United States, wind turbines resulted in the deaths of 888,000 bats and 573,000 birds, including 83,000 birds of prey.

Also in 2013, a meta-analysis by Scott Loss and others in the journal found that the likely mean number of birds killed annually in the U.S by monopole tower wind turbines was 234,000. The authors acknowledged the larger number reported by Smallwood, but noted that Smallwood's meta-analysis did not distinguish between types of wind turbine towers. The monopole towers used almost exclusively for new wind installations have mortality rates that "increase with increasing height of monopole turbines", but as of yet, it remains to be determined if increasingly taller monopole towers result in lower mortality per GWh.

Bird mortality at wind energy facilities can vary greatly depending on the location, construction, and height, with some facilities reporting zero bird fatalities, and others as high as 9.33 birds per turbine per year.

A comprehensive study of wind turbine bird deaths by the Canadian Wildlife Service in 2013 analyzed reports from 43 out of the 135 wind farms operating across Canada as of December 2011. After adjusting for search inefficiencies, the study found an average of 8.2 bird deaths per tower per year, from which they arrived at a total of 23,000 per year for Canada at that time. Actual habitat loss averaged 1.23 hectares per turbine, which involved the direct loss of, on average, 1.9 nesting sites per turbine. The effective habitat loss, which was not quantified, was observed to be highly variable between species: some species avoided nesting within 100 to 200 m from turbines, while other species were observed feeding on the ground directly under the blades. The study concluded that, overall, the combined effect on birds was "relatively small" compared to other causes of bird mortality, but noted that mitigation measures might be required in some situations to protect at-risk species.

While studies show that other sources, such as cats, cars, buildings, power lines, and transmission towers kill far more birds than wind turbines, many studies and conservation groups have noted that wind turbines disproportionately kill large migratory birds and birds of prey, and are more likely to kill birds threatened with extinction. Wind facilities have attracted the most attention for impacts on iconic raptor species, including golden eagles. The Pine Tree Wind energy project near Tehachapi, California has one of the highest raptor mortality rates in the country; by 2012 at least eight golden eagles had been killed according to the U.S. Fish and Wildlife Service (USFWS). Biologists have noted that it is more important to avoid losses of large birds as they have lower breeding rates and can be more severely impacted by wind turbines in certain areas.

Large numbers of bird deaths are also attributed to collisions with buildings. An estimated 1 to 9 million birds are killed every year by tall buildings in Toronto, Ontario, Canada alone, according to the wildlife conservation organization Fatal Light Awareness Program. Other studies have stated that 57 million are killed by cars, and some 365 to

988 million are killed by collisions with buildings and plate glass in the United States alone. Promotional event lightbeams as well as ceilometers used at airport weather offices can be particularly deadly for birds, as birds become caught in their lightbeams and suffer exhaustion and collisions with other birds. In the worst recorded ceilometer lightbeam kill-off during one night in 1954, approximately 50,000 birds from 53 different species died at the Warner Robins Air Force Base in the United States.

Arctic terns and a wind turbine at the Eider Barrage in Germany.

In the United Kingdom, the Royal Society for the Protection of Birds (RSPB) concluded that "The available evidence suggests that appropriately positioned wind farms do not pose a significant hazard for birds." It notes that climate change poses a much more significant threat to wildlife, and therefore supports wind farms and other forms of renewable energy as a way to mitigate future damage. In 2009 the RSPB warned that "numbers of several breeding birds of high conservation concern are reduced close to wind turbines" probably because "birds may use areas close to the turbines less often than would be expected, potentially reducing the wildlife carrying capacity of an area.

Concerns have been expressed that wind turbines at Smøla, Norway are having a deleterious effect on the population of white-tailed eagles, Europe's largest bird of prey. They have been the subject of an extensive re-introduction programme in Scotland, which could be jeopardised by the expansion of wind turbines.

The Peñascal Wind Power Project in Texas is located in the middle of a major bird migration route, and the wind farm uses avian radar originally developed for NASA and the United States Air Force to detect birds as far as 4 miles (6.4 km) away. If the system determines that the birds are in danger of running into the rotating blades, the turbines shut down and are restarted when the birds have passed. A 2005 Danish study used surveillance radar to track migrating birds traveling around and through an offshore wind farm. Less than 1% of migrating birds passing through an offshore wind farm in Rønde, Denmark, got close enough to be at risk of collision, though the site was studied only during low-wind conditions. The study suggests that migrating birds may avoid large turbines, at least in the low-wind conditions the research was conducted in. Furthermore, it is not thought that nocturnal migrants are at higher risk to collision than diurnally active species.

In the above figure old style wind turbines at Altamont Pass in California, which are being replaced by more "bird-friendly designs". While newer designs are taller, there is as yet, no definitive evidence that they are "friendlier". A recent study suggests that they might not be safer to wildlife, and are not a "simple fix", according to Oklahoma State University ecologist Scott Loss.

In 2012, researchers reported that, based on their four-year radar tracking study of birds after construction of an offshore wind farm near Lincolnshire, that pink-footed geese migrating to the U.K. to overwinter altered their flight path to avoid the turbines.

At the Altamont Pass Wind Farm in California, a settlement between the Audubon Society, Californians for Renewable Energy and NextEra Energy Resources who operate some 5,000 turbines in the area requires the latter to replace nearly half of the smaller turbines with newer, more bird-friendly models by 2015 and provide $2.5 million for raptor habitat restoration. The proposed Chokecherry and Sierra Madre Wind Energy Project in Wyoming is allowed by the Bureau of Land Management (BLM) to "take" up to 16 eagles per year as predicted by the Fish and Wildlife Service, while making power-lines less damaging. A 2012 BLM study estimated nearly 5,400 birds each year, including over 150 raptors. Some sites are required to watch for birds. In 2016, the Obama administration finalized a rule that granted 30-year licenses to wind-energy companies that operate high-speed turbines permitting them to kill or injure up to 4,200 golden eagles and bald eagles, four times the existing limit, before facing penalties. There are 143,000 bald eagles and 40,000 golden eagles in the United States.

Bats

Bats may be injured by direct impact with turbine blades, towers, or transmission lines. Recent research shows that bats may also be killed when suddenly passing through a low air pressure region surrounding the turbine blade tips.

The numbers of bats killed by existing onshore and near-shore facilities have troubled bat enthusiasts.

In April 2009 the Bats and Wind Energy Cooperative released initial study results showing a 73% drop in bat fatalities when wind farm operations are stopped during low wind conditions, when bats are most active. Bats avoid radar transmitters, and placing microwave transmitters on wind turbine towers may reduce the number of bat collisions.

A 2013 study produced an estimate that wind turbines killed more than 600,000 bats in the U.S. the previous year, with the greatest mortality occurring in the Appalachian Mountains. Some earlier studies had produced estimates of between 33,000 and 888,000 bat deaths per year.

Weather and Climate Change

Wind farms may affect weather in their immediate vicinity. This turbulence from spinning wind turbine rotors increases vertical mixing of heat and water vapor that affects the meteorological conditions downwind, including rainfall. Overall, wind farms lead to a slight warming at night and a slight cooling during the day time. This effect can be reduced by using more efficient rotors or placing wind farms in regions with high natural turbulence. Warming at night could "benefit agriculture by decreasing frost damage and extending the growing season. Many farmers already do this with air circulators".

A number of studies have used climate models to study the effect of extremely large wind farms. One study reports simulations that show detectable changes in global climate for very high wind farm usage, on the order of 10% of the world's land area. Wind power has a negligible effect on global mean surface temperature, and it would deliver "enormous global benefits by reducing emissions of CO_2 and air pollutants". Another peer-reviewed study suggested that using wind turbines to meet 10 percent of global energy demand in 2100 could actually have a warming effect, causing temperatures to rise by 1 °C (1.8 °F) in the regions on land where the wind farms are installed, including a smaller increase in areas beyond those regions. This is due to the effect of wind turbines on both horizontal and vertical atmospheric circulation. Whilst turbines installed in water would have a cooling effect, the net impact on global surface temperatures would be an increase of 0.15 °C (0.27 °F). Author Ron Prinn cautioned against interpreting the study "as an argument against wind power, urging that it be used to guide future research". "We're not pessimistic about wind," he said. "We haven't absolutely proven this effect.

Impacts on People

Aesthetics

Aesthetic considerations of wind power stations have often a significant role in their evaluation process. To some, the perceived aesthetic aspects of wind power stations may conflict with the protection of historical sites. Wind power stations are less likely to be perceived negatively in urbanized and industrial regions. Aesthetic issues are subjective and some people find wind farms pleasant or see them as symbols of energy independence and local prosperity. While studies in Scotland predict wind farms will damage tourism, in other countries some wind farms have themselves become tourist attractions, with several having visitor centers at ground level or even observation decks atop turbine towers.

The surroundings of Mont Saint-Michel at low tide. While windy coasts are good locations for wind farms, aesthetic considerations may preclude such developments in order to preserve historic views of cultural sites.

In the 1980s, wind energy was being discussed as part of a soft energy path. Renewable energy commercialization led to an increasing industrial image of wind power, which is being criticized by various stakeholders in the planning process, including nature protection associations. Newer wind farms have larger, more widely spaced turbines, and have a less cluttered appearance than older installations. Wind farms are often built on land that has already been impacted by land clearing and they coexist easily with other land uses.

Coastal areas and areas of higher altitude such as ridgelines are considered prime for wind farms, due to constant wind speeds. However, both locations tend to be areas of high visual impact and can be a contributing factor in local communities' resistance to some projects. Both the proximity to densely populated areas and the necessary wind speeds make coastal locations ideal for wind farms.

Wind power stations can impact on important sight relations which are a key part of culturally important landscapes, such as in the Rhine Gorge or Moselle valley. Conflicts between heritage status of certain areas and wind power projects have arisen in various countries. In 2011 UNESCO raised concerns regarding a proposed wind farm 17 kilometres away from the French island abbey of Mont-Saint-Michel. In Germany, the impact of wind farms on valuable cultural landscapes has implications on zoning and land-use planning. For example, sensitive parts of the Moselle valley and the background of the Hambach Castle, according to the plans of the state government, will be kept free of wind turbines.

Loreley rock in Rhineland-Palatinate, part of UNESCO World heritage site Rhine Gorge.

Wind turbines require aircraft warning lights, which may create light pollution. Complaints about these lights have caused the US FAA to consider allowing fewer lights per turbine in certain areas. Residents near turbines may complain of "shadow flicker" caused by rotating turbine blades, when the sun passes behind the turbine. This can be avoided by locating the wind farm to avoid unacceptable shadow flicker, or by turning the turbine off for the time of the day when the sun is at the angle that causes flicker. If a turbine is poorly sited and adjacent to many homes, the duration of shadow flicker on a neighbourhood can last hours.

Wind Turbine Syndrome

Wind turbine syndrome is a psychosomatic disorder largely caused by anxiety about wind farms and not by the turbines themselves. There is limited evidence of anxiety effects caused by low level noise in the close vicinity of the turbines.

Safety

Some turbine nacelle fires cannot be extinguished because of their height, and are sometimes left to burn themselves out. In such cases they generate toxic fumes and can cause secondary fires below. Newer wind turbines, however, are built with automatic fire extinguishing systems similar to those provided for jet aircraft engines. These autonomous systems, which can be retrofitted to older wind turbines, automatically detect a fire, shut down the turbine unit, and extinguish the fires.

During winter, ice may form on turbine blades and subsequently be thrown off during operation. This is a potential safety hazard, and has led to localised shut-downs of turbines. A 2007 study noted that no insurance claims had been filed, either in Europe or the US, for injuries from ice falling from wind towers, and that while some fatal accidents have occurred to industry workers, only one wind-tower related fatality was known to occur to a non-industry person: a parachutist.

Given the increasing size of production wind turbines, blade failures are increasingly relevant when assessing public safety risks from wind turbines. The most common failure is the loss of a blade or part thereof

Offshore

Many offshore wind farms have contributed to electricity needs in Europe and Asia for years, and as of 2014 the first offshore wind farms are under development in U.S. waters. While the offshore wind industry has grown dramatically over the last several decades, especially in Europe, there is still some uncertainty associated with how the construction and operation of these wind farms affect marine animals and the marine environment.

Traditional offshore wind turbines are attached to the seabed in shallower waters within the near-shore marine environment. As offshore wind technologies become more advanced, floating structures have begun to be used in deeper waters where more wind resources exist.

Common environmental concerns associated with offshore wind developments include:

- The risk to seabirds being struck by wind turbine blades or being displaced from critical habitats;

- Underwater noise associated with the installation process of monopile turbines;

- The physical presence of offshore wind farms altering the behavior of marine mammals, fish, and seabirds by reasons of either attraction or avoidance;

- Potential disruption of the near-field and far-field marine environments from large offshore wind projects.

Germany restricts underwater noise during pile driving to less than 160 dB.

Due to the landscape protection status of large areas of the Wadden Sea, a major World Heritage Site with various national parks (e.g. Lower Saxon Wadden Sea National Park) German offshore installations are mostly restricted on areas outside the territorial waters. Offshore capacity in Germany is therefore way behind the British or Danish near coast installments, which face much lower restrictions.

In January 2009, a comprehensive government environmental study of coastal waters in the United Kingdom concluded that there is scope for between 5,000 and 7,000 offshore wind turbines to be installed without an adverse impact on the marine environment. The study—which forms part of the Department of Energy and Climate Change's Offshore Energy Strategic Environmental Assessment—is based on more than a year's research. It included analysis of seabed geology, as well as surveys of sea birds and marine mammals. There does not seem to have been much consideration however of the likely impact of displacement of fishing activities from traditional fishing grounds.

ENVIRONMENTAL IMPACTS OF HYDROELECTRIC POWER

Hydroelectric power includes both massive hydroelectric dams and small run-of-the-river plants. Large-scale hydroelectric dams continue to be built in many parts of the world (including China and Brazil), but it is unlikely that new facilities will be added to the existing U.S. fleet in the future.

Instead, the future of hydroelectric power in the United States will likely involve increased capacity at current dams and new run-of-the-river projects. There are environmental impacts at both types of plants.

Land Use

The size of the reservoir created by a hydroelectric project can vary widely, depending largely on the size of the hydroelectric generators and the topography of the land. Hydroelectric plants in flat areas tend to require much more land than those in hilly areas or canyons where deeper reservoirs can hold more volume of water in a smaller space.

At one extreme, the large Balbina hydroelectric plant, which was built in a flat area of Brazil, flooded 2,360 square kilometers—an area the size of Delaware—and it only provides 250 MW of power generating capacity (equal to more than 2,000 acres per

MW) . In contrast, a small 10 MW run-of-the-river plant in a hilly location can use as little 2.5 acres (equal to a quarter of an acre per MW).

Flooding land for a hydroelectric reservoir has an extreme environmental impact: it destroys forest, wildlife habitat, agricultural land, and scenic lands. In many instances, such as the Three Gorges Dam in China, entire communities have also had to be relocated to make way for reservoirs .

Wildlife Impacts

Dammed reservoirs are used for multiple purposes, such as agricultural irrigation, flood control, and recreation, so not all wildlife impacts associated with dams can be directly attributed to hydroelectric power. However, hydroelectric facilities can still have a major impact on aquatic ecosystems. For example, though there are a variety of methods to minimize the impact (including fish ladders and in-take screens), fish and other organisms can be injured and killed by turbine blades.

Apart from direct contact, there can also be wildlife impacts both within the dammed reservoirs and downstream from the facility. Reservoir water is usually more stagnant than normal river water. As a result, the reservoir will have higher than normal amounts of sediments and nutrients, which can cultivate an excess of algae and other aquatic weeds. These weeds can crowd out other river animal and plant-life, and they must be controlled through manual harvesting or by introducing fish that eat these plants. In addition, water is lost through evaporation in dammed reservoirs at a much higher rate than in flowing rivers.

In addition, if too much water is stored behind the reservoir, segments of the river downstream from the reservoir can dry out. Thus, most hydroelectric operators are required to release a minimum amount of water at certain times of year. If not released appropriately, water levels downstream will drop and animal and plant life can be harmed. In addition, reservoir water is typically low in dissolved oxygen and colder than normal river water. When this water is released, it could have negative impacts on downstream plants and animals. To mitigate these impacts, aerating turbines can be installed to increase dissolved oxygen and multi-level water intakes can help ensure that water released from the reservoir comes from all levels of the reservoir, rather than just the bottom (which is the coldest and has the lowest dissolved oxygen).

Life-Cycle Global Warming Emissions

 Global warming emissions are produced during the installation and dismantling of hydroelectric power plants, but recent research suggests that emissions during a facility's operation can also be significant. Such emissions vary greatly depending on the size of the reservoir and the nature of the land that was flooded by the reservoir.

Small run-of-the-river plants emit between 0.01 and 0.03 pounds of carbon dioxide equivalent per kilowatt-hour. Life-cycle emissions from large-scale hydroelectric plants built in semi-arid regions are also modest: approximately 0.06 pounds of carbon dioxide equivalent per kilowatt-hour. However, estimates for life-cycle global warming emissions from hydroelectric plants built in tropical areas or temperate peatlands are much higher. After the area is flooded, the vegetation and soil in these areas decomposes and releases both carbon dioxide and methane. The exact amount of emissions depends greatly on site-specific characteristics. However, current estimates suggest that life-cycle emissions can be over 0.5 pounds of carbon dioxide equivalent per kilowatt-hour.

To put this into context, estimates of life-cycle global warming emissions for natural gas generated electricity are between 0.6 and 2 pounds of carbon dioxide equivalent per kilowatt-hour and estimates for coal-generated electricity are 1.4 and 3.6 pounds of carbon dioxide equivalent per kilowatt-hour.

ENVIRONMENTAL IMPACTS OF GEOTHERMAL ENERGY

The most widely developed type of geothermal power plant (known as hydrothermal plants) are located near geologic "hot spots" where hot molten rock is close to the earth's crust and produces hot water. In other regions enhanced geothermal systems (or hot dry rock geothermal), which involve drilling into Earth's surface to reach deeper geothermal resources, can allow broader access to geothermal energy.

Geothermal plants also differ in terms of the technology they use to convert the resource to electricity (direct steam, flash, or binary) and the type of cooling technology they use (water-cooled and air-cooled). Environmental impacts will differ depending on the conversion and cooling technology used.

Water Quality and Use

Geothermal power plants can have impacts on both water quality and consumption. Hot water pumped from underground reservoirs often contains high levels of sulfur, salt, and other minerals. Most geothermal facilities have closed-loop water systems, in which extracted water is pumped directly back into the geothermal reservoir after it has been used for heat or electricity production. In such systems, the water is contained within steel well casings cemented to the surrounding rock. There have been no reported cases of water contamination from geothermal sites in the United States.

Water is also used by geothermal plants for cooling and re-injection. All U.S. geothermal power facilities use wet-recirculating technology with cooling towers. Depending on the cooling technology used, geothermal plants can require between 1,700 and 4,000 gallons of water per megawatt-hour. However, most geothermal plants can use either geothermal fluid or freshwater for cooling; the use of geothermal fluids rather than freshwater clearly reduces the plants overall water impact.

Most geothermal plants re-inject water into the reservoir after it has been used to prevent contamination and land subsidence. In most cases, however, not all water removed from the reservoir is re-injected because some is lost as steam. In order to maintain a constant volume of water in the reservoir, outside water must be used. The amount of water needed depends on the size of the plant and the technology used; however, because reservoir water is "dirty," it is often not necessary to use clean water for this purpose. For example, the Geysers geothermal site in California injects non-potable treated wastewater into its geothermal reservoir.

Air Emissions

The distinction between open- and closed-loop systems is important with respect to air emissions. In closed-loop systems, gases removed from the well are not exposed to the atmosphere and are injected back into the ground after giving up their heat, so air emissions are minimal. In contrast, open-loop systems emit hydrogen sulfide, carbon dioxide, ammonia, methane, and boron. Hydrogen sulfide, which has a distinctive "rotten egg" smell, is the most common emission.

Once in the atmosphere, hydrogen sulfide changes into sulfur dioxide (SO_2). This contributes to the formation of small acidic particulates that can be absorbed by the bloodstream and cause heart and lung disease. Sulfur dioxide also causes acid rain, which damages crops, forests, and soils, and acidifies lakes and streams. However, SO_2 emissions from geothermal plants are approximately 30 times lower per megawatt-hour than from coal plants, which is the nation's largest SO_2 source.

Some geothermal plants also produce small amounts of mercury emissions, which must be mitigated using mercury filter technology. Scrubbers can reduce air emissions, but they produce a watery sludge composed of the captured materials, including sulfur, vanadium, silica compounds, chlorides, arsenic, mercury, nickel, and other heavy metals. This toxic sludge often must be disposed of at hazardous waste sites.

Land Use

The amount of land required by a geothermal plant varies depending on the properties of the resource reservoir, the amount of power capacity, the type of energy conversion system, the type of cooling system, the arrangement of wells and piping systems, and the substation and auxiliary building needs. The Geysers, the largest geothermal plant in the world, has a capacity of approximately 1,517 megawatts and the area of the plant is approximately 78 square kilometers, which translates to approximately 13 acres per megawatt. Like the Geysers, many geothermal sites are located in remote and sensitive ecological areas, so project developers must take this into account in their planning processes.

Land subsidence, a phenomenon in which the land surface sinks, is sometimes caused by the removal of water from geothermal reservoirs. Most geothermal facilities address this risk by re-injecting wastewater back into geothermal reservoirs after the water's heat has been captured.

Hydrothermal plants are sited on geological "hot spots," which tend to have higher levels of earthquake risk. There is evidence that hydrothermal plants can lead to an even greater earthquake frequency. Enhanced geothermal systems (hot dry rock) can also increase the risk of small earthquakes. In this process, water is pumped at high pressures to fracture underground hot rock reservoirs similar to technology used in natural gas hydraulic fracturing. Earthquake risk associated with enhanced geothermal systems can be minimized by siting plants an appropriate distance away from major fault lines. When a geothermal system is sited near a heavily populated area, constant monitoring and transparent communication with local communities is also necessary.

Life-Cycle Global Warming Emissions

In open-loop geothermal systems, approximately 10 percent of the air emissions are carbon dioxide, and a smaller amount of emissions are methane, a more potent global warming gas. Estimates of global warming emissions for open-loop systems are approximately 0.1 pounds of carbon dioxide equivalent per kilowatt-hour. In closed-loop systems, these gases are not released into the atmosphere, but there are a still some emissions associated with plant construction and surrounding infrastructure.

Enhanced geothermal systems, which require energy to drill and pump water into hot rock reservoirs, have life-cycle global warming emission of approximately 0.2 pounds of carbon dioxide equivalent per kilowatt-hour.

To put this into context, estimates of life-cycle global warming emissions for natural gas generated electricity are between 0.6 and 2 pounds of carbon dioxide equivalent per kilowatt-hour and estimates for coal-generated electricity are 1.4 and 3.6 pounds of carbon dioxide equivalent per kilowatt-hour.

References

- Dolan, Stacey L.; Heath, Garvin A. (2012). "Life Cycle Greenhouse Gas Emissions of Utility-Scale Wind Power". Journal of Industrial Ecology. 16: S136–S154. Doi:10.1111/j.1530-9290.2012.00464.x. SSRN 2051326

- Environmental-impacts-geothermal-energy, renewable-energy, our-energy-choices, clean-energy: ucsusa.org Retrieved 13 July, 2019

- Johns, Robert. Actions by Feds Cut Annual Bird Deaths in Oil and Gas Fields by Half, Saving Over One Million Birds From Grisly Death, Washington, D.C.: American Bird Conservancy, January 3, 2013. Retrieved July 30, 2013

- Impact-of-solar-energy-on-the-environment: greenmatch.co.uk, Retrieved 24 March, 2019

- "Wind energy Frequently Asked Questions". British Wind Energy Association. Archived from the original on 2006-04-19. Retrieved 2006-04-21

- Environmental-impacts-hydroelectric-power, renewable-energy, our-energy-choices, clean-energy: ucsusa.org, Retrieved 1 June, 2019

Permissions

All chapters in this book are published with permission under the Creative Commons Attribution Share Alike License or equivalent. Every chapter published in this book has been scrutinized by our experts. Their significance has been extensively debated. The topics covered herein carry significant information for a comprehensive understanding. They may even be implemented as practical applications or may be referred to as a beginning point for further studies.

We would like to thank the editorial team for lending their expertise to make the book truly unique. They have played a crucial role in the development of this book. Without their invaluable contributions this book wouldn't have been possible. They have made vital efforts to compile up to date information on the varied aspects of this subject to make this book a valuable addition to the collection of many professionals and students.

This book was conceptualized with the vision of imparting up-to-date and integrated information in this field. To ensure the same, a matchless editorial board was set up. Every individual on the board went through rigorous rounds of assessment to prove their worth. After which they invested a large part of their time researching and compiling the most relevant data for our readers.

The editorial board has been involved in producing this book since its inception. They have spent rigorous hours researching and exploring the diverse topics which have resulted in the successful publishing of this book. They have passed on their knowledge of decades through this book. To expedite this challenging task, the publisher supported the team at every step. A small team of assistant editors was also appointed to further simplify the editing procedure and attain best results for the readers.

Apart from the editorial board, the designing team has also invested a significant amount of their time in understanding the subject and creating the most relevant covers. They scrutinized every image to scout for the most suitable representation of the subject and create an appropriate cover for the book.

The publishing team has been an ardent support to the editorial, designing and production team. Their endless efforts to recruit the best for this project, has resulted in the accomplishment of this book. They are a veteran in the field of academics and their pool of knowledge is as vast as their experience in printing. Their expertise and guidance has proved useful at every step. Their uncompromising quality standards have made this book an exceptional effort. Their encouragement from time to time has been an inspiration for everyone.

The publisher and the editorial board hope that this book will prove to be a valuable piece of knowledge for students, practitioners and scholars across the globe.

Index